Famous Maritime Squadrons of the RAF

Volume 1

by James J Halley

Illustrations by
Michael Trim
Thomas Brittain
Terry Hadler
David Palmer

General Editor
Charles W Cain

Hylton Lacy Publishers Limited, Windsor, Berkshire, England

© Hylton Lacy Publishers Limited, 1973

First published in England 1973
by Hylton Lacy Publishers Limited
Coburg House Sheet Street Windsor
Berkshire England

First Edition 1973
ISBN 0 85064 101 2

Series Editor Charles W Cain

Available as Men and Machines books

Famous Combat Units of the World's
Air Forces series
Famous Fighter Squadrons of the RAF   Volume I
Carrier Air Groups : HMS *Eagle*

Aircraft of World War Two series
American Fighters	Volume I
German Air Force Fighters	Volume I
Royal Air Force Bombers	Volume I
German Air Force Bombers	Volume I
Royal Air Force Fighters	Volume I
Japanese Navy Bombers	
German Air Force Fighters	Volume II
Royal Air Force Bombers	Volume II
German Air Force Bombers	Volume II
American Fighters	Volume II
French Fighters	
American Bombers	Volume I
French Bombers (in preparation)	

Printed in England by Chichester Press Ltd, Chichester, Sussex

Foreword

In this second volume of histories of famous Royal Air Force squadrons, five units have been selected which have served mainly in a maritime role. Three are typical of the relatively small number of flying-boat squadrons whose personnel formed a band of specialists in marine aviation set aside from the rest of the Service.

For forty years the 'Flying-boat Union' developed and maintained an expertize in the many facets of operating water-based aircraft. The sight of their stately boats in full flight was to remain an indelible memory for those fortunate enough to have watched them.

Of the remaining two squadrons, one became famous as a torpedo-bomber unit—perhaps the most hazardous of all wartime tasks. The other was representative of the land-based patrol squadrons whose original role of coastal reconnaissance soon gave place to lengthy sorties over the oceans. Between them, these five squadrons have covered the entire range of Coastal Command duties at one time or another.

As before, the space available for each squadron's history is restricted and certain aspects have been given more detailed treatment than others. Periods during which the squadrons were engaged in roles other than maritime have been passed over in favour of those during which a squadron was involved in more naval activities.

Four of the squadrons are currently active, two with maritime reconnaissance Nimrods and two with air-sea rescue Whirlwinds. The fifth, No. 204 Squadron, disbanded some time before this volume was completed.

In dealing with the operations of these squadrons, it is inevitable that, of the units, it is the aircrews that are mentioned in the text. But behind every flight lay a vast amount of hard work and organization by ground crews, especially in maritime squadrons. Maintaining flying-boats on wintry, wind-swept anchorages was no sinecure and, more than most, maritime aircraft were dependent on good maintenance and accurate briefings.

In World War Two, RAF Coastal Command operations seldom received the attention of the Press accorded daily to RAF Fighter and Bomber Commands. Many Coastal operations ended, after long hours in bad weather over an inhospitable ocean, with no obvious result. At most, and for many crews, an attack on a U-boat was an infrequent occurrence. There was no way of knowing that their often lone presence over a convoy had averted an attack and saved ships and seamen. Once their stint of patrolling was completed, they then faced a long trip home to base and a landing, as often as not, in darkness or rain on a stretch of water dotted with obstructions. More RAF Coastal aircraft were lost through bad weather and accidents than by enemy action but without the far-reaching patrols the Battle of the Atlantic could never have been won.

The main sources of information on each squadron's history have been the contemporary Operations Record Books maintained since shortly before World War Two as a day-to-day record of activities. We are very grateful for the unfailing assistance of the Air Historical Branch, Ministry of Defence, in producing from its archives the relevant documents and to the staff of the Public Record Office in whose care are kept the records of World War One and the years between the wars.

We would also like to thank the personnel of the Imperial War Museum and the photographic libraries of the Ministry of Defence and *Flight International*. The publications of Air-Britain (The International Association of Aviation Historians) and the IPMS (International Plastic Modellers Society) have been invaluable in providing background information on maritime aircraft and their colour schemes.

Our appreciation must also be expressed to the members of the five selected squadrons for their courtesy and assistance during visits and correspondence: to Squadron Leader J. Weaver and Flight Lieutenant C. G. Goodman of No. 22 Squadron; Wing Commander J. M. Alcock and Flying Officer A. J. Lovett of No. 201 Squadron; Squadron Leader R. G. Reekie and Flying Officer M. Buckley of No. 202 Squadron; Squadron Leaders D. E. Leppard and K. J. Miles of No. 204 Squadron; and Wing Commander J. Wild and Squadron Leader S. C. Hall of No. 206 Squadron.

Not least we are indebted to Mr Norman B. Wiltshire for his aid in photographing Squadron Standards and the preparation of the colour plates for this series.

James J. Halley
Shepperton, Middlesex, England
June 1972

Contents

Introduction	9
No. 22 Squadron	11
No. 22 Squadron—Aircraft, Bases, Commanders	26
No. 201 Squadron	27
No. 201 Squadron—Aircraft, Bases, Commanders	43
No. 202 Squadron	44
No. 202 Squadron—Aircraft, Bases, Commanders	57
No. 204 Squadron	58
No. 204 Squadron—Aircraft, Bases, Commanders	68
No. 206 Squadron	69
No. 206 Squadron—Aircraft, Bases, Commanders	84

Colour Illustrations

Squadron Badges	5
Maps of Squadron Bases: *United Kingdom*	6
Mediterranean	7
Squadron Standards	8
No. 22 Squadron	18
No. 201 Squadron	34
No. 202 Squadron	46
No. 204 Squadron	66
No. 206 Squadron	78

The illustrations on the facing page are from copies of the original drawings prepared by The College of Arms. The badges incorporate the crown used at the time the badges were approved for these squadrons, in all cases the King George VI crown. If carried on current aircraft, they would incorporate the Queen's crown. The badges are reproduced with the permission of the Ministry of Defence and the Chester Herald (The Inspector of Royal Air Force Badges) and with the kind assistance of the Royal Air Force Club.

No. 22 Squadron: On a *torteau*, a Maltese cross throughout. Over all a 'pi' fimbriated. Motto: *'Preux et audacieux'*—'Valiant and brave'.

No. 201 Squadron: A seagull, wings elevated and addorsed. Motto: *'Hic et ubique'*—'Here and everywhere'.

No. 202 Squadron: A mallard alighting. Motto: *'Semper vigilate'*—'Be always vigilant'.

No. 204 Squadron: On water barry wavy, a mooring buoy, thereon a cormorant displayed. Motto: *'Praedam mari quaero'*—'I seek my prey in the sea'.

No. 206 Squadron: An octopus. Motto: *'Nihil nos effugit'*—'Naught escapes us'.

© Hylton Lacy Publishers Ltd

#	Name	#	Name	#	Name	#	Name	#	Name
1	Culli Voe	29	Martlesham Heath	57	Carew Cheriton	85	De Kooy	113	Pola
2	Sullom Voe	30	Felixstowe	58	Pembroke Dock	86	Den Helder	114	Taranto
3	Lerwick	31	Harwich	59	Valley	87	Emden	115	Naples
4	Sumburgh	32	Eastchurch	60	Hooten Park	88	Jever	116	Spezia
5	Hatston	33	Chatham	61	Sealand	89	Wilhelmshaven	117	Toulon
6	Scapa Flow	34	Westgate	62	Silloth	90	Nordholz	118	Genoa
7	Stornoway	35	Manston	63	Wig Bay	91	Stade	119	Ajaccio
8	Wick	36	Walmer	64	Stranraer	92	Finkenwerder		
9	Tain	37	Dover	65	Prestwick	93	Kiel		
10	Alness	38	Detling	66	Bowmore	94	Sylt		
11	Invergordon	39	Tangmere	67	Tiree	95	Aalborg		
12	Lossiemouth	40	Ford	68	Oban	96	Kristiansand/Kjevik		
13	Kinloss	41	Thorney Island	69	Benbecula	97	Horten		
14	Dyce	42	Gosport	70	Castle Archdale	98	Kjeller		
15	Tayport	43	Portsmouth	71	Londonderry	99	Fornebu		
16	Leuchars	44	Calshot	72	Ballykelly	100	Mandal		
17	Donibristle	45	Lee-on-Solent	73	Aldergrove	101	Lista		
18	Rosyth	46	Portland	74	Cherbourg	102	Stavanger		
19	East Fortune	47	Exeter	75	Abbeville	103	Vaernes		
20	Thornaby	48	Mount Batten	76	Petite Synthe	104	Gibraltar		
21	Driffield	49	Plymouth	77	Coudekerque	105	Arzew		
22	Leconfield	50	Falmouth	78	Bray Dunes	106	Oran		
23	Killingholme	51	Portreath	79	Varssenaere	107	Bizerta		
24	North Coates	52	St. Mawgan	80	Haamstede	108	Malta		
25	Donna Nook	53	St. Eval	81	Waalhaven	109	Alexandria		
26	Bircham Newton	54	Davidstowe Moor	82	Valkenburg	110	Mirabella		
27	Coltishall	55	Chivenor	83	Schiphol	111	Kotor		
28	Oakington	56	Lyneham	84	Schellingwoude	112	Split		

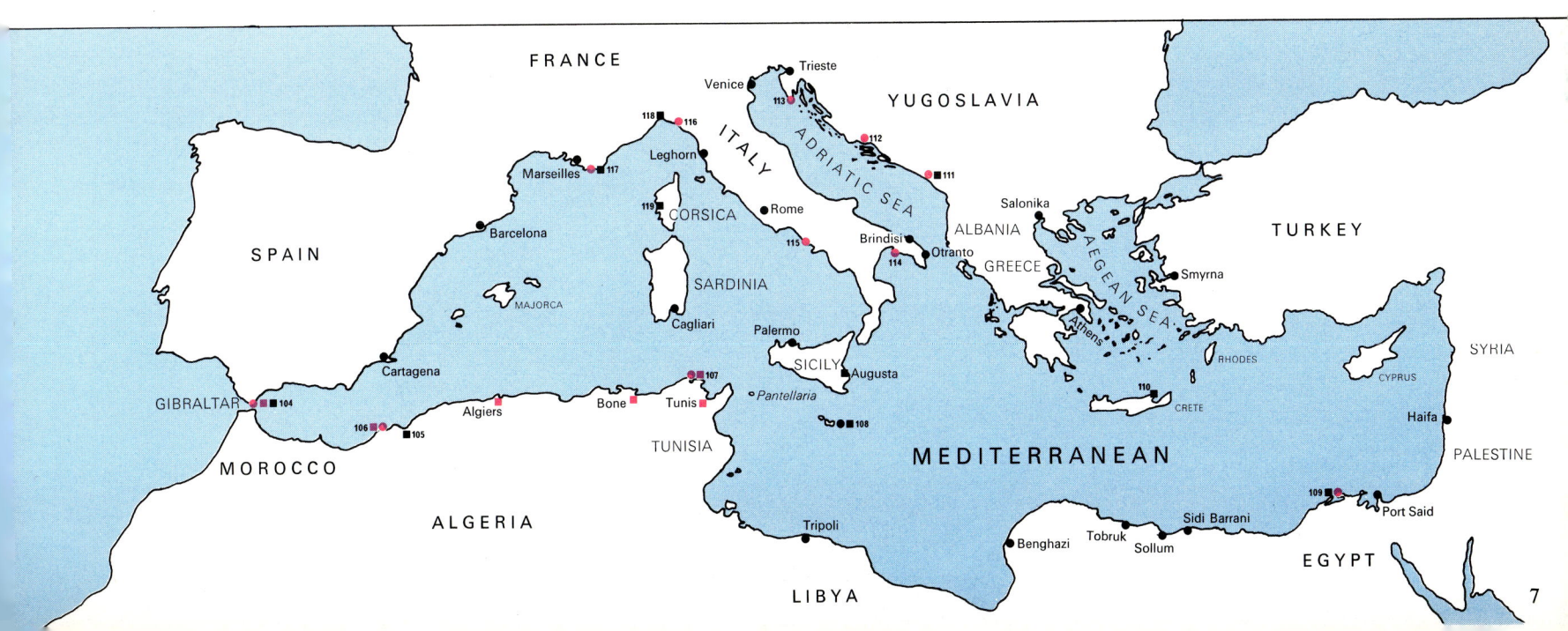

No. 22 Squadron

SQUADRON STANDARDS

The Squadron Standards of the five Squadrons covered by this volume are reproduced on this page. Each carries individual battle honours on scrolls flanking the Squadron Badge as follows:

World War One

Arras	No. 201 and 206
Hindenburg Line	No. 22
Somme	No. 22 and 201
Western Front	No. 22, 201, 202, 204 and 206
Ypres	No. 22, 201

World War Two

Arctic	No. 204
Atlantic	No. 201, 202, 204 and 206
Biscay	No. 202 and 206
Bismarck	No. 201 and 206
Burma	No. 22
Channel and North Sea	No. 22 and 206
Dunkirk	No. 206
Eastern Waters	No. 22
Fortress Europe	No. 206
Home Waters	No. 204
Mediterranean	No. 22 and 202
Normandy	No. 201
North Africa	No. 202
Norway	No. 201 and 204

No. 201 Squadron

No. 204 Squadron

No. 206 Squadron

No. 202 Squadron

© Hylton Lacy Publishers Limited

Introduction

A nation which is an island must, of necessity, command the seas around it if its security is to be maintained. For centuries, Britain depended on its Navy but exactly three centuries after these islands became a united kingdom, a new element was introduced by man's first brief foray into the air in a powered, heavier-than-air machine. Only six years later Monsieur Louis Blériot flew across the English Channel.

The development of aircraft was rapid. From faltering hops in a dead calm, aeroplanes increased in speed and range; airships could cruise for many hours and were looked upon as possible replacements for cruisers as the eyes of a modern navy. In a decade, a revolution in naval tactics had taken place, not only over but under the sea. Submarines had emerged from their dangerous experimental period and were now an accepted class of warship. Two mutually-antagonistic components of sea power had arrived on the scene simultaneously.

When World War One erupted in August 1914, the Royal Navy had undergone a complete change of tactical doctrine and equipment since the end of the 19th century. Steam-powered ships of iron and steel had freed warships from the restrictions of wind and tide. The launching of the *Dreadnought* in 1905 began a new building programme of 'all big gun' battleships which rendered earlier classes obsolete. Unfortunately, these leviathans were vulnerable to attack from lurking submarines against which there was no defence other than ramming.

The menace of the submarine resulted in the deployment of the British Grand Fleet to Scapa Flow, in the Orkney Islands, where it could intercept any foray by the German High Seas Fleet into the Atlantic. It could also cut off any force attacking the East Coast but both tasks required good reconnaissance. This the Germans possessed in their fleet of long-ranging rigid airships while the Royal Navy relied on seaplanes operating from seaplane carriers and very dependent on calm seas. More seaworthy craft were available in the shape of flying-boats based at a few coastal stations of the Royal Naval Air Service.

As the war progressed, sorties by the German Fleet became remote, especially after its hair's-breadth escape at the Battle of Jutland. In contrast, the menace of the U-boats increased month by month. Attacks on merchant ships of all nationalities became general and until a convoy system was established, losses were heavy. By 1917, there was extreme danger of starvation facing the population of the British Isles. Fortunately, anti-submarine weapons were becoming effective and ships equipped with sound detectors and depth charges began to take a toll of the U-boats.

As escorts for convoys, destroyers were the best shepherds but the number available was limited. The Grand Fleet required the best for its screen. Those that could be spared were supplemented by sloops and submarine-chasers. Also they had the aid of the RNAS with their airships and seaplanes. While the latter lacked the ability to sink U-boats except in very favourable circumstances, the very presence of observers in the sky was a deterrent to submarine commanders who had an instinctive aversion to being watched.

To provide additional protection around the coasts, numerous land bases were established and were manned by flights of Airco D.H.6 trainers. These two-seat biplanes were never designed for over-water reconnaissance, their unreliable engines being responsible for depositing more than one crew in the sea. At full bore, they would permit the carriage of an observer to spot a U-boat or a bomb to drop on it—but seldom both.

Several coastal stations had begun to operate large flying-boats by the end of the war. Designed by Squadron Commander John Porte, RNAS, the Felixstowe-series (the 'F-boats') had been put into production in 1917 to replace a handful of imported Curtiss boats operating from Felixstowe and Yarmouth. To these bases were added Calshot, Cattewater, Tresco in the Scilly Isles, Killingholme and Dundee; while six flights were allocated to the Orkneys and Shetlands as support for the Grand Fleet.

With a duration of up to six hours, the F-boats extended the patrol area out over the Western Approaches and North Sea. The oceans were vast but, inevitably, shipping funnelled into the English Channel, Bristol Channel and to the north of Ireland. Where merchant ships massed, U-boats were to be found seeking their prey. Augmented—during the final months of the war—by US Navy flying-boats based in Ireland and Brittany, patrols criss-crossed the shipping lanes; but submarine detection was, invariably, a matter of keen and practised eyesight.

As the war ended, experiments were bearing fruit with hydroplanes mounted on seaplanes and flying-boats which could be operated from waterborne aircraft. Nevertheless, the 'Mark One Eyeball' remained the sole submarine detection device for more than two decades and is still an invaluable aid in spotting submarines. The four observation blisters on an electronic-packed Nimrod bears witness to this irreplaceable component.

The Armistice on 11 November 1918 brought to an end many developments in maritime aviation. A vast maritime air force was rapidly dismantled, its aircraft scrapped, sold or stored, its personnel demobilized. From the chain of coastal stations, only a handful remained—Calshot and Cattewater (later renamed Mount Batten) by their proximity to the naval bases at Portsmouth and Plymouth, together with Felixstowe as a result of its experimental role. Overseas, anti-submarine aircraft had congregated in the Mediterranean and the seaplane station at Calafrana on Malta had become a major base while the nearby dockyard built Felixstowe F.3 flying-boats with local labour. On the heel of Italy, a force of aircraft patrolled over the Strait of Otranto and other bases were sprinkled along the convoy routes to Egypt and the Aegean. By the middle of 1920, these had contracted to a single station at Malta.

The diminutive peacetime Royal Air Force had little scope for large flying-boats. During the 1920s, there were usually more experimental than operational boats on strength. Even the three operational units (two at home and one in Malta) were reduced from squadrons to flights. A score of wartime F.5s kept alive the anti-submarine and maritime reconnaissance capability of the coastal units.

After testing a variety of prototypes, the Air Ministry ordered the Supermarine Southampton as the RAF's standard flying-boat for the late 1920s. Coming into service in 1925, the type laid the foundations for an expansion of the flying-boat units in 1929 and 1930. Flights became squadrons and new squadrons were formed. Squadrons appeared at Singapore and in the Persian Gulf.*

A second type of maritime unit came into being with the formation of land-based torpedo-bomber units for Coast Defence duties. The first two formed were transferred to Singapore as part of the defences of the new naval base and were replaced by two new squadrons.

Though two-motor types formed the basis of the RAF's flying-boat squadrons, a few tri-motor aircraft were provided; the Short Rangoon was a modified civil airliner but the Blackburn Iris and Perth were large military boats. Southamptons were replaced by Supermarine Scapas and, later, Saro Londons and Supermarine Stranraers. Supplementing these were the majestic four-motor Short Singapore IIIs. One of the benefits bestowed by the new boats was the ability to reach Malta from Gibraltar and Egypt without refuelling. As Gibraltar was also within reach of Mount Batten, this meant that flying-boats could be deployed to the Middle and Far East without touching foreign territory.

In 1935, an expansion scheme for the RAF began. The old Coastal Area became RAF Coastal Command in 1936 and new squadrons began to form for general reconnaissance (GR) duties. Their equipment was the Avro Anson, the RAF's first monoplane with a retractable undercarriage to enter squadron service. To augment the flying-boat units, two new types were developed. Both were monoplanes, one twin-engined and the other four-engined. The first was the Saro Lerwick which proved a complete failure while the second was destined to be the last and greatest of all RAF flying-boats—the Short Sunderland.

When World War Two broke out in September 1939, only a small number of Sunderlands was in service, aided by obsolescent biplane Londons, Stranraers and Singapores. Ansons, already out-dated, were being replaced by Hudsons—a military development of the civil Lockheed Model 14—as fast as these more modern aircraft could be built and brought across the Atlantic. Blackburn Bothas built for general reconnaissance proved to be unreliable and were relegated to training duties. Armstrong Whitworth Whitleys and, later Vickers-Armstrong Wellington bombers were diverted to Coastal Command and were reinforced from across the Atlantic by Consolidated Catalinas and Liberators and, after a disastrous debut as a day bomber, the Boeing Fortress.

For the first half of the war, aircraft failed to sink many U-boats because of the lack of effective anti-submarine weapons and search radar. They did, however, keep U-boats submerged and located others while escorting convoys, enabling surface craft to forestall attacks on their charges. A combination of adequate radar, improved depth charges and hard-won experience resulted in an increasing number of decisive encounters.

By the end of 1944, air attack had become the major cause of losses to U-boats. To recharge storage batteries, the German submarines were forced to surface for lengthy periods but during the final year 'schnorkels' began to be fitted. These were ventilation tubes which could be raised above the surface to enable diesel engines to be operated while the U-boat was still submerged. Only the tip of the '*Schnorkel*', and a trail of diesel fumes, marked the position of the U-boat. Precision radar, and sonobuoys or listening devices which were dropped in a pattern to locate submerged submarines, enabled some successes to be recorded against 'schnorkeling' submarines before the end of the war.

By 1945, a vast network of patrol bases stretched around the world. Royal Air Force Coastal Command patrols from Great Britain, Iceland and the Azores met in mid-Atlantic their opposite numbers of the Royal Canadian Air Force and the United States Navy from North America. Strike wings operated against enemy shipping off the Norwegian coast and in the Baltic. In the Mediterranean, similar units had virtually exhausted the supply of suitable targets; but numbers of patrol squadrons ensured the security of the convoy routes supplying the Allied armies in Italy and the Balkans.

Other maritime squadrons were based in East and West Africa, while India and Ceylon housed flying-boat and shore-based Liberator units to counter the Imperial Japanese Navy in the Indian Ocean. Inexorably, Japanese coastal shipping in the Bay of Bengal was brought to a halt by air attack and minelaying operations.

With the end of the war, many maritime units were disbanded, leaving a small number of Sunderland squadrons in the United Kingdom and Far East while land-based squadrons re-equipped with Avro Lancasters To replace the wartime Lancasters, a new maritime-reconnaissance aircraft was developed; the Avro (later Hawker Siddeley) Shackleton entered service in 1951 and remains in service over 20 years later. Strike aircraft were abandoned by RAF Coastal Command after the war; this task being given to bomber and naval aircraft.

By the end of the 1950s flying-boats were phased-out and Shackletons became the sole maritime-reconnaissance type, some Lockheed Neptunes having been acquired as gap-fillers for a few years. As a replacement, the de Havilland (later Hawker Siddeley) Comet airliner was developed into a maritime aircraft under the name of Nimrod and entered service with the squadrons in 1970.

Royal Air Force Coastal Command was disbanded on 27 November 1969 and in its place came No. 18 (Maritime) Group of RAF Strike Command. At the beginning of 1972, the maritime force had contracted to four Nimrod squadrons (Nos. 42, 120, 201 and 206) and an operational conversion unit (No. 236) in the UK and one in the Mediterranean (No. 203). Two Shackleton squadrons were left, one (No. 204) for shipping reconnaissance and one (No. 8) reformed with early-warning Shackletons while two Westland Whirlwind helicopter squadrons (Nos. 22 and 202) provided search and rescue services around the coasts of Britain. In Cyprus, No. 84 Squadron was engaged in similar tasks. Between them, these squadrons formed the remnants of the RAF's once-powerful maritime command.

---

*Today, this stretch of water is occasionally referred to by certain unilateral oil-conscious interests as 'The Arabian Gulf' and by others, with multilateral interests but equal historic disregard, as 'The Gulf'—EDITOR.

22 Squadron

F.E.2b (serial A5216) at Bertangles shows its early gun mounting in the gunner's cockpit and a nose wheel to prevent nosing-over, the latter protecting the exposed crew in their flimsy nacelle (Photo: IWM Q63183)

Fighters in France

For the greater part of its service with the Royal Air Force, No. 22 Squadron has been connected with maritime operations. In common with many other famous units, No. 22 Squadron started life in quite a different role; and, as it began to form on 1 September 1915, its immediate destiny within the Royal Flying Corps lay in France.

The obsolete forts on the edge of Gosport aerodrome were part of the historic landward defences of the great naval base of Portsmouth, Hampshire. Long since however, these have been used as accommodation for numerous units. From a few rooms in Fort Grange, No. 22 Squadron was soon evicted to Fort Rowner, a half-mile further north, where a nucleus of personnel grew as officers and men arrived to join those supplied by No. 13 Squadron. Before the end of September 1915, 'A' Flight had received some B.E.2cs and Maurice Farman Shorthorns and 'B' Flight had acquired a number of Bleriots. Besides personnel, No. 13 Squadron had also passed on three Bleriots, a Caudron G.III and a B.E. 8a. To these were added a variety of other types including Curtiss J.N.3s, Martinsyde S.1A and Bristol Scouts and Avro 504As. Not before February 1916 did the squadron receive its first operational aircraft—F.E.2b two-seat reconnaissance/escort fighters.

Until the establishment of full-time Home Defence squadrons, aircraft were supplied from training units to fly defensive patrols around London. That No. 22 was to engage in anti-Zeppelin flights became clear with the arrival in October of eight boxes of 'Darts, RAF, No. 3', the 'RAF' referring to the contemporary Royal Aircraft Factory* at Farnborough, Hampshire. These were to be dropped on any marauding airship—presuming that the defending aircraft could climb above it. Few did and No. 22 never sighted an enemy even to attempt the feat.

The collection of miscellaneous types used for training at this time was a common sight on British airfields. As squadrons received their operational aircraft, so they passed on the remnants of their training equipment—as No. 13 Squadron had to No. 22. For example, a Bleriot Parasol (military serial number 570) had survived several squadron's attempts to dismember it before ending its career in No. 22 Squadron by catching fire after landing at Gosport on 16 November 1915, and being reduced to ashes. Then again, a Curtiss J.N.3 (No. 6117) which had arrived from No. 2 Reserve Aircraft Park at Northolt, Middlesex, in November, crashed on landing on 8 December and was soon 'cannibalized' to provide parts for other aircraft.

On a different tack, when the polished aluminium cowlings of the B.E.2cs affected visibility by their reflections, the squadron painted them dull khaki.

By the end of 1915, mechanics had to cope with two types of Bleriot, Curtiss J.N.3s, B.E.2cs, Martinsyde S.1As and Bristol Scouts; with the added aero-engine complications of 50 h.p. and 80 h.p. Gnomes, 90 h.p. Curtiss' and 70 h.p. Renaults. Twenty-one aircraft were flyable and nine under repair.

On 12 February 1916, the squadron's first F.E.2b

*With the creation of the Royal Air Force on 1 April 1918, 'The Factory' was renamed the Royal Aircraft Establishment, a title which lives on today in the 1970s.

(No. 5216) arrived from the Norfolk works of Boulton & Paul Ltd at Norwich; followed closely by another 'Fee' (No. 5217)—which was wrecked 16 days later. Re-equipment with a battle-worthy aircraft was under way and the squadron's B.E.s—no fewer than 14—were handed over to No. 33 Squadron. The Squadron's Blériots, six in number, were ordered to be sent to London, to Gamage Bells Garage, Ebury Bridge Road, Pimlico. However, they were then diverted to the Thames Ironworks at Greenwich; an equally unlikely address for aircraft but indicative of the variety of establishments becoming involved in aviation.

Finally, on 15 March 1916, the squadron's transport left Gosport for France to join the British Expeditionary Force and on 1 April, Captain G. R. Howard led a formation of 12 'Fees' to France, passing through St Omer on its way to Vert Galand near Doullens to join the 14th (Army) Wing. Three other F.E.2b squadrons had already arrived on the Western Front and soon after moving to Bertangles, just north of St Omer, No. 22 began to fly reconnaissance missions over the trench lines.

Compared with earlier types, the F.E.2b was a far more effective combat aircraft than the Royal Flying Corps had flown in the previous 18 months. Derived from the F.E.2a designed and built by the Royal Aircraft Factory, the F.E.2b was a two-seater powered by a 120 h.p. Beardmore pusher engine, the tail assembly being carried by four stout struts. The nacelle accommodated the pilot immediately in front of the Beardmore and a lower cockpit for the observer also housed the armament; initially a single 0·303-in Lewis gun on a

pillar mounting. When firing, the observer was restrained from leaving the aircraft by a strap anchored to the cockpit but was completely exposed above the thighs. At such times, the armour sheet fitted to the bottom of the nacelle was for moral support only.

Early in May 1916, No. 22 began to increase its strength from 12 to 18 aircraft and completed this by mid-June. By this time, the first Sopwith 1½ Strutters had begun to arrive in France and were the first to be fitted with interruptor gear enabling a forward-firing machine-gun to fire through the propeller arc. In contrast, German single-seat fighters with synchronized machine-guns had appeared in the summer of 1915 and in the absence of comparable Allied equipment pusher-engine designs had been sent to France to counter them. Though intended for reconnaissance, the F.E.2b made an ideal gun platform and squadrons equipped with the type found themselves engaged more and more in escort duties. The bulky engine provided a 'blind-spot' for enemy fighters and many 'Fees' carried an additional Lewis gun on a pillar capable of firing rearwards over the top wing.

To improve performance, the F.E.2b was allocated the 160 h.p. Beardmore and supplies of these reached No. 22 by the end of 1916. Despite its ungainly appearance, the 'Fee' was comparatively agile and the squadron flew fighter and reconnaissance missions until July 1917.

Experience of air warfare during 1915 had indicated a clear requirement for a replacement for the slow-moving B.E.2 variants that equipped the bulk of the Royal Flying Corps. Among the designs under development in the spring of 1916 was one by the British & Colonial Aeroplane Company (Bristol) for a two-seat reconnaissance biplane powered by a Rolls-Royce engine. This was modified to a fighter-reconnaissance role and in September, 1916, it appeared as the Bristol F.2A. Armed with a fixed 0·303-in Vickers machine-gun for the pilot and a flexible 0·303-in Lewis gun for the observer, the type proved to be fast and manoeuvrable, its engine destined to become the famous Rolls-Royce Falcon. Fifty aircraft were ordered and in March 1917, No. 48 Squadron arrived in France equipped with the first operational Bristol F.2As. On its first patrol over the Front, six of the squadron's aircraft met a patrol of enemy single-seat fighters and only two returned. This inauspicious beginning was due to the tactics initially employed. When attacked by fighters, a formation of two-seaters maintained close formation to permit their gunners to concentrate their fire on any attacker.

It was soon discovered that the new two-seater was quite capable of taking care of itself if flown in the manner of a single-seat scout but with the gunner covering the usual blind-spot behind the tail. This made

1 Fort Grange, Gosport, in 1918, shows the casemates of the old fort and, behind, the sheds on the airfield. Among the trees in the background is Fort Brockhurst (*Photo: Imperial War Museum ref. Q80654*)

2 Bristol Fighters of 'B' Flight taking-off over the hangars at Serny and parked Camels of No. 208 Squadron in June 1918. Each aircraft carried an individual recognition letter for easy identification in the air (*Photo: IWM. Q10331*)

3 Bristol Fighter of No. 22 Squadron prepares for take-off at Vert Galand on 1 April 1918, the day on which the Royal Air Force was formed (*Photo: IWM. Q11997*)

4 A heavily-armed Bristol Fighter with an unusual top wing mounting for a Lewis 0·303-in. gun. Below the observer's cockpit is the warning '*This machine must not be flown without passenger or equivalent ballast in gunner's cockpit*', below which are 23 'iron crosses' to indicate enemy aircraft shot down (*Photo: IWM. Q69691*)

the Bristol Fighter highly unpopular with the enemy and for the rest of the war the Bristol two-seat fighter was rated by them as one of their most dangerous opponents.

No. 22 received Bristol F.2Bs in July 1917—an improved version of the original batch—and aircrews found the difference in performance between the 'Biff' and its 'Fee' predecessor startling. By 24 August of the same year the squadron had its full complement of 18 aircraft and long offensive patrols over the Lines began to be flown. The toll of enemy aircraft grew, culminating in the destruction of 84 during May 1918 alone.

When forced to withdraw from its airfield by the German offensive in March 1918, No. 22 flew from fields farther west until October. As the German army retired, the squadron moved east behind the front line, the end of the war finding it near Douai. For the next six months, No. 22 was based at Nivelles in Belgium before moving to Spich, nine miles from Cologne, as part of the Army of Occupation. At the end of August 1919, No. 22 Squadron left for home, leaving its faithful 'Biffs' behind. It remained as a cadre unit until the end of the year, demobilization having robbed No. 22 of most of its personnel.

Testing Time

On 24 July 1923, No. 22 Squadron was reformed at Martlesham Heath near Ipswich, Suffolk. This airfield was the base of A. & A.E.E.—the Aeroplane and Armament Experimental Establishment. Two squadrons, Nos. 15 and 22, were formed to test aircraft and armament for the Establishment, No. 22 being given the task of evaluating prototypes offered for possible service with the Royal Air Force. A large number of types passed through the squadron's hands during the next 10 years; and not without risk to its pilots. Several fatal accidents were suffered, notably involving the Gloster Gamecock and Blackburn Turcock—the latter being a lightweight fighter for the Turkish Air Force which crashed on 23 January 1928. The adoption of parachutes by the RAF saved the lives of a number of No. 22's pilots. One of them was Flying Officer A. J. Pegg who stepped out of an Avro Tutor (K3191) on 21 March 1933 when its wing broke-up in a dive. Later, he became famous for his test flying over many years at the Bristol Aeroplane Company, culminating in the giant Brabazon and the turboprop Britannia.

In 1934, it was decided that the A. & A.E.E. would be responsible for its own test flying and both testing squadrons were disbanded, No. 22 leaving Martlesham on 1 May 1934.

The Tin Fish Men

On the same day as No. 22 Squadron ceased to exist at Martlesham Heath, a new Squadron came into being in Scotland; at Donibristle on the north shore of the Firth of Forth. During the previous year, No. 100 Squadron had left for the Far East to join No. 36 Squadron at Singapore. Both were torpedo-bomber units and No. 100's departure left no torpedo-bomber squadron operational in the British Isles, the only aircraft of this type being used by the Coast Defence Development Flight at Gosport which was involved in experimental work.

To fill this gap, No. 22 was designated a torpedo-bomber squadron and received six of the first batch of Vildebeest Is delivered by Vickers in 1932. These were large two-seat biplanes powered by a single 600 h.p. Bristol Pegasus radial engine permitting a maximum speed of about 140mph. Though slow, the type had a strong steel-tube airframe and good manoeuvrability—both assets in an aircraft whose role was the delivery of a 2150-lb torpedo at low altitudes. The torpedo's weight was also destined to increase over the years.

Torpedo-carrying aircraft were not new. Both the British and German air arms had made successful torpedo attacks during World War One but found that the weight of the naval torpedos used restricted the performance of the carrier seaplanes to a point where flight was barely possible. Surprise enabled initial attacks to succeed but more efficient torpedo-carriers were required to make sustained operations possible. In Britain, the Sopwith Cuckoo and Short Shirl single-seat torpedo-bombers were developed for naval use, the former being built in numbers for a projected attack on the German High Seas Fleet in its harbours. Training took place at East Fortune, across the Firth of Forth from Donibristle, and Cuckoos were embarked in HMS *Argus* but the war ended before the operation could be mounted.

In the decade following the 1918 Armistice, development of torpedo aircraft was entirely based on carrier-borne operations. Blackburn Darts were embarked in carriers and developed many of the techniques required for successful attacks. The decision to base torpedo-bombers ashore grew from a requirement for aircraft to defend the naval base which was being built at Singapore. Fortunately, Hawker Horsley day bombers already in service with the RAF could be modified to carry torpedoes and, in June 1928, No. 36 Squadron received the first of these. By the end of 1930, the squadron had moved to the Far East; to be followed by No. 100 which had been converted to a torpedo-bomber unit at the same time.

Under the command of Squadron Leader T. A. Warne-Browne, No. 22 Squadron began practising with its new equipment. The first requirement was the ability to drop a torpedo from the right height, at the right speed and in the right position. This was not at all simple; the torpedo had to enter the water at the correct angle, otherwise, either it would bounce back into the air or dive to the bottom of the sea. Dropping at too great a speed could cause the torpedo to break its back. With too little speed, the aircraft was liable to accompany its torpedo into the water. Finally, unless the correct angle was computed for the torpedo run, the torpedo would miss the target altogether—and torpedo-bombers had no 'second-strike' capability.

On 18 September 1934, No. 22 began carrying out dummy attacks on naval vessels in the Firth of Forth. Ten days later, a series of attacks on the Home Fleet resulted in three 'hits' on the battle-cruiser HMS *Renown*, two more on the light cruiser *Leander* and one each on the aircraft carrier *Courageous* and the battle-cruiser *Hood*, the last-mentioned being the then largest naval target in the world. A score of 100% was assisted by the slow speed and lack of evasive action displayed by the ships of the Fleet. But there was no doubt remaining that torpedo attack from the air was destined to be a potential menace to ships at sea.

Later trials were more difficult. A light cruiser was provided as a target and, in 12 attacks on HMS *Curacao*, six hits were attained. When during the next attack, the cruiser took avoiding action the score dropped to three out of 12. Subsequently a change in the weather resulted in no hits from the next four attacks and the remaining eight aircraft were recalled.

Optimum results relied on a combination of various circumstances. Knowledge of the location of the target was essential in view of the short range of the Vildebeest. As the enemy would be unlikely to steer a steady course while under attack, tactics were evolved which enabled a spread of torpedoes to be launched so that in evading one torpedo the ship would cross the track of another. Getting several aircraft into their correct dropping positions at precisely the right time was a problem not made simpler by sea mist or bad weather. On the run-up an attacking aircraft would be subjected to anti-aircraft fire from the target and any other accompanying ships. Fortunately, the slow speed of the bomber was in part compensated for by the prevailing low standard of naval AA—anti-aircraft—gunnery.

Attrition while flying close to the sea was higher than in other types of squadron. Out of about 45 Vildebeest allotted to No 22, no fewer than 15 were written-off in accidents. This was despite the rugged structure of the type which kept the Vildebeest (and its GP or general-purpose variant, the Vincent) in service until 1944 and on operational flying until March 1942.

In March 1935, the squadron had begun to re-equip with Vildebeest IIIs. This mark, with a more powerful Pegasus, had accommodation for a third crew member; an observer being carried to ease navigational problems. A full complement of these aircraft were taken to the

---

1 'Dos and Don'ts' for torpedo-bomber pilots are shown in an Air Ministry instructional diagram which depicts the Royal Air Force's first torpedo-bomber, the Sopwith Cuckoo (*Photo: IWM. Q67846*)

2 This Vildebeest (K2822) was an elderly Mark I which saw service with No. 22 for the first two months of World War Two (*Photo: No. 22 Squadron Collection*)

3 Vildebeest I (S1710) flying over Fareham, Hampshire, during a flight from Gosport, home of the Torpedo Development Unit. The first Vildebeest mark to be issued to No. 22 Squadron was this two-seat version (*Photo: IWM. MH25*)

Mediterranean in October 1935, when a large number of RAF squadrons was deployed during the Italian attack on Ethiopia. No. 22 was posted to Hal Far, Malta, where it could cover most of the Central Mediterranean.

Embarking on the troopship *Cameronian* at Glasgow on 3 October, the squadron arrived on the 10th and had its aircraft erected and ready for operations on the 21st. It was at Hal Far that No. 22 received its official badge which incorporated a Maltese Cross and replaced an unofficial badge in use since July 1934 which was the family crest of the Earl of Moray and the motto *Semper resurgam* ('I shall always rise again').

On 20 July 1936, preparations for a return to the UK were ordered and four days later all aircraft were despatched by road to Kalafrana seaplane station for crating. Squadron personnel embarked in the troopship *Somersetshire* on 21 August and arrived at Southampton on 29 August. After a period of leave, a detachment was sent to Sealand, Cheshire, which erected the squadron's aircraft. They were then flown to Donibristle, Scotland.

'B' Flight was detached on 14 December 1936 to form No. 42 Squadron, accompanied by Flight Lieutenant W. G. Campbell who returned to No. 22 in April 1937 to become its Commanding Officer. There were now two torpedo-bomber squadrons in the UK, both armed with Vildebeest. However, design work was in progress on its successor, the Bristol Beaufort, which was to make its first flight on 15 October 1938; an order for 78 having been already placed.

Torpedo training continued with the Home Fleet. Occasionally the opportunity arose when dummy attacks could be made on a major force, as in July 1937 when the squadron moved to Exeter Airport for a few days to take part in a coastal defence exercise. Attacks off Cornwall on the carriers HMS *Furious* and *Courageous* and the battleships *Ramillies* and *Revenge*

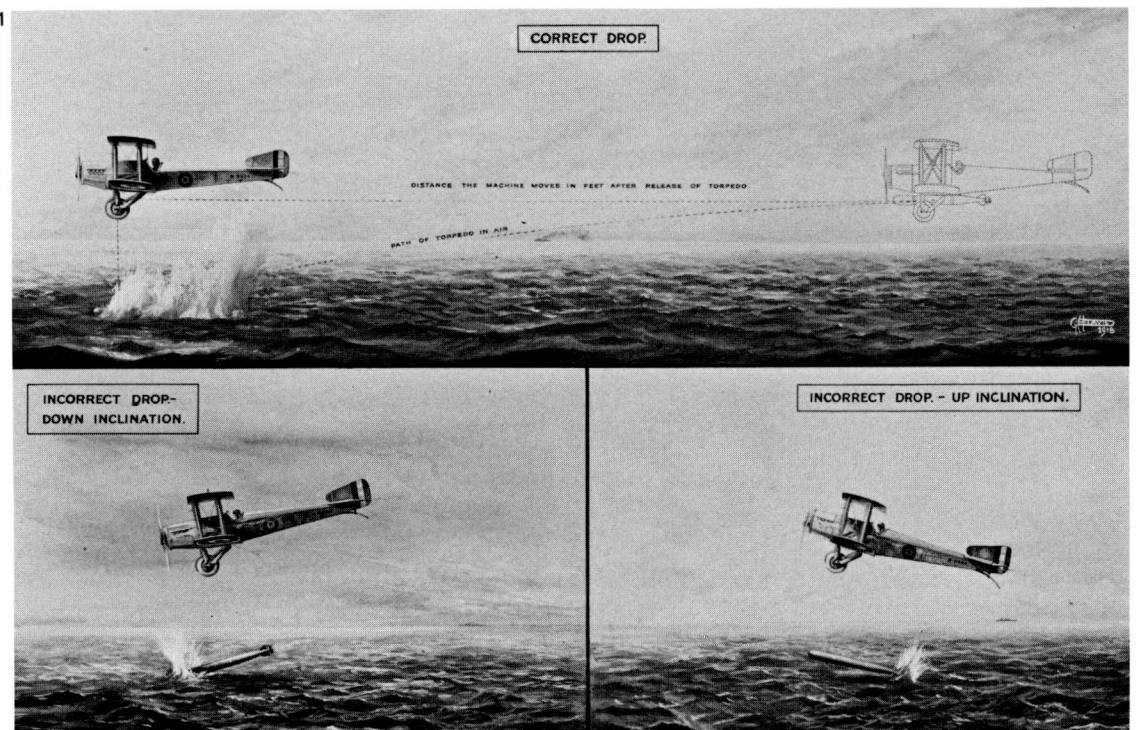

resulted in five hits on the carriers; umpires assessed casualties to the torpedo-bombers as 14 aircraft. On the final day, a Vildebeest flight surprised *Furious* south of Exeter and scored three hits before they could be engaged by defensive AA 'fire'.

In March 1938, the squadron changed base to the new airfield at Thorney Island, Sussex, a short distance along the coast from Portsmouth. Here they joined the Vildebeest IVs of No. 42 Squadron and their envy at the latter's later Mk IVs was tempered by the knowledge that No. 22 would be the first to receive the new Beauforts.

The Munich Crisis in October 1938 resulted in the addition of camouflage paint to both the shore-base station and the aircraft. However, another year of uneasy peace lay ahead. When it expired, both squadrons were still flying the Vildebeest Mks III and IV; and, No. 42 with its more modern (*sic*) aircraft departed for Bircham Newton in Norfolk, while No. 22's pilots began twin-engine conversion courses on Avro Ansons and Bristol Blenheims. In the meantime, patrols began on 19 September over the English Channel carrying six 100-lb anti-submarine bombs.

On 15 November 1939, the squadron's first Beaufort flew into Thorney Island followed by three more before the end of the month. The new torpedo-bomber was a handsome two-motor monoplane powered by 1010 h.p. Bristol Taurus air-cooled, two-row radials. With a top

15

speed of 265 m.p.h. and a range of over 1000 miles, the new type was a vast improvement on the obsolete Vildebeest. The power-operated dorsal turret was supposed to be fitted with twin-mounted 0·303-in machine-guns but, at first, only single guns were available. Later a rearward-firing twin-gun blister was fitted under the nose. Since, under operational conditions, it was not particularly effective, it was often removed. Beam guns were fitted to augment the turret guns and, in place of the nose under gun position, one or two 0·303-in Vickers K-guns were mounted on gimbals. The 18-in torpedo was stowed in a recess under the fuselage to reduce drag.

As already related, torpedo-bombing was not a science easy to learn. Much training was necessary to convert Vildebeest crews to their new mounts. Teething troubles with the new Taurus radials resulted in several engine failures. Three Beauforts were lost during January 1940 from this cause; fortunately without casualties. The Vildebeest (by now Mark IVs) still carried on patrolling through some of the worst weather ever experienced over the English Channel and when HM King George VI travelled to France in December 1940, his ship was provided with an escort of four of No. 22's venerable biplanes. On 20 December, K4591 flew the squadron's last operational mission with the faithful Vildebeest.

On 8 April 1940, no fewer than 14 Beauforts flew up to North Coates in Lincolnshire at 2000 feet under solid cloud as No. 22 Squadron changed its base. With the German invasion of Norway and Denmark under way, both Nos. 22 and 42 were put on stand-by. However, the latter squadron was in the throes of Vildebeest 'disposal' and conversion to Beauforts was in full swing. On 15 April, nine aircraft were despatched to lay mines off the mouth of the Elbe. One (L4465) failed to return.

Mining (code-name '*Gardening*') was a major task for Beauforts in the summer of 1940. Mines ('*Vegetables*') were planted in areas with horticultural names (for example, '*Nectarines*') which eventually ranged from Norway and the Baltic to the Spanish coast. Such sorties were normally flown at night, not only for protection from enemy fighters but also to conceal the spot where the mines were laid. Torpedoes were on hand for attacks on enemy warships but appearances of German heavy ships were infrequent and fleeting. Twelve aircraft were despatched to Lossiemouth in Morayshire, on stand-by for attacks on enemy battleships and cruisers off Norway; but were not used. On 7 May, six aircraft left to attack an enemy cruiser reported between Norderney and Juist in the Friesian Islands. Three aircraft bombed the ship (which was probably a destroyer) and a Beaufort (coded 'G/22'; serial L4464) was reported missing with Flying Officer Woollatt and his crew. Wing Commander Mellor's Beaufort; (C/22 L4518) returned damaged with a wounded gunner and had to be crash-landed. On the same day, another Beaufort (L4466) crashed on the beach adjacent to the airfield—Pilot Officer Berryman was killed as yet another victim of engine failure at a crucial moment. Two days later, four out of five minelaying Beauforts were forced to return by bad weather; the fifth (L4453) was not heard from again.

The German invasion of the Low Countries on 10 May brought the squadron to a state of 1-hour stand-by; and next day eight aircraft were detailed for mining. On 12 May, six Beauforts set out on their first anti-shipping strike but were recalled when 20 miles from the Dutch coast. Five more dropped two 500-lb bombs each on Waalhaven Airport, Rotterdam, which had been captured by German parachute troops; they returned safely. On 18 May, six Beauforts joined Lockheed Hudsons of No. 233 Squadron in a night attack on an oil storage depot in Hamburg but only one identified and bombed the target.

The Commanding Officer failed to return from a minelaying sortie on 25 May (in 'F/22'; L4450); and, four days later, five aircraft carried out the squadron's first daylight attack on shipping—the target being motor torpedo-boats in IJmuiden harbour. During the vital days of the Battle of France, bombing raids were made on targets in Belgium. Two more Beauforts were lost through engine failure and a Court of Enquiry was set up to investigate the problem.

Nine aircraft were sent to Wick on 12 June and next day, seven set out to attack Bergen but lost formation in bad visibility and were forced to turn back. A second attempt on the 15th was undertaken by eight Beauforts led by Wing Commander Braithwaite (in 'J/22') who attacked an oil tanker in Sognefjord. Other Beauforts bombed a steamer, an ammunition dump and three other objectives near Bergen without loss. An attempted strike by three aircraft on the battleship *Scharnhorst* in

Trondheim-fjord was forced to return by bad weather and the opportunity did not recur.

During July, Beauforts were grounded for modifications and the squadron began training at Gosport with 'Toraplanes'. This was a type of winged torpedo invented by Sir Dennis Burney which consisted of a structure attached to an ordinary torpedo to enable it to be released from any height. The gyroscope of the torpedo would be connected to the controls to enable the Toraplane to glide at a one-in-seven angle at 160 m.p.h. into the water. A complementary weapon was the 'Doraplane' which was a more sophisticated winged missile designed to fly horizontally at a predetermined height for up to one mile. Both weapons were early forms of 'stand-off' missiles which would enable aircraft to attack ships and harbours from outside AA range.

At a meeting in May 1939, chaired by Air Vice-Marshall Arthur Tedder, (then Director-General of Research and Development), it was agreed that tests should be carried out on 'Toraplanes'. Messrs Kryn and Lahy of Letchworth built a prototype which consisted of a set of wings connected by a spar along the top of the torpedo to the tail assembly. The structure was released close to the water by a paravane trailing behind and the torpedo entered the water alone. The Admiralty supplied a Fairey Swordfish (L2831) for fitting trials at Henlow and Beaufort torpedo crutches were sent to Kryn and Lahy when the initial model could not be attached to the Swordfish due to various parts fouling the structure of the aircraft. During July, satisfactory release tests were carried out at Gosport and in August full-scale dropping trials in Stokes Bay took place. The 'Toraplane' glided for 750 yards and entered the water safely after being released from a Swordfish crossing the area in formation with a Short Singapore flying-boat, a Vickers Wellesley landplane bomber and another Swordfish—all carrying observers.

Despite its original promise, the 'Toraplane' never became operational with No. 22 Squadron, notwithstanding much time having been devoted to training. While this was taking place, US-built Martin Marylands —the first for the RAF—began to arrive as No. 22 had been instructed to train a nucleus of a new reconnaissance squadron with these aircraft. These two-motor bombers had been originally ordered by the French as the Martin 167A-3. After France surrendered, the remainder of the contract was taken over by Britain and the Model 167F (company designation) was given the name of Maryland.

On the last day of August, operations were resumed when 'G/22' (L9790) and 'Q/22' (L9791) were sent to bomb the Schellingwoude seaplane station in Amsterdam harbour. Barges assembled in enemy-occupied ports in preparation for an invasion of Britain were bombed in early September at the height of the Battle of Britain. Then, on 11 September, the squadron finally got a chance to use its primary weapons.

Flight Lieutenant Dick Beauman led a formation of five Beauforts armed with torpedoes to attack a convoy off Calais. Failing to rendezvous with their fighter escort over Detling in Kent, the Beauforts set out for the reported position of the enemy ships. Nothing was found so course was set eastwards and, off Ostend, several ships were sighted. On the run-in, three torpedoes failed to release and a fourth hit a sandbank and exploded. But the one Beauman dropped ran true and hit a 6000-ton ship while the other Beauforts raked the escort vessels to subdue enemy defensive *flak*. All five aircraft returned safely from this first strike with torpedoes.

---

1 The Vildebeest III had an additional cockpit fitted behind the pilot. Wheel spats were not normally carried on Mark IIIs in order to avoid mud clogging the wheels on grass airfields. Slots are fitted to the leading-edge of the top wing to improve slow speed performance (*Photo: IWM. MH24*)

2 A trio of early production Beauforts running-up. In addition to the squadron code (OA) and individual aircraft letters fore and aft of the roundel, No. 22's Beauforts carried the aircraft letter on the nose (*Photo: IWM. CH643*)

3 Beauforts and crews at North Coates. Aircrew are wearing 'Mae West' flotation jackets and parachute harness ; the parachute packs being stowed until required in contrast to fighter pilots' seat-type parachutes (*Photo: IWM. CH647*)

4 Beaufort 'F/22' (W6537) flies low over the sea, its camouflage merging with the grey Northern waters. The dorsal turret is armed with two guns (*Photo: IWM. CH7492*)

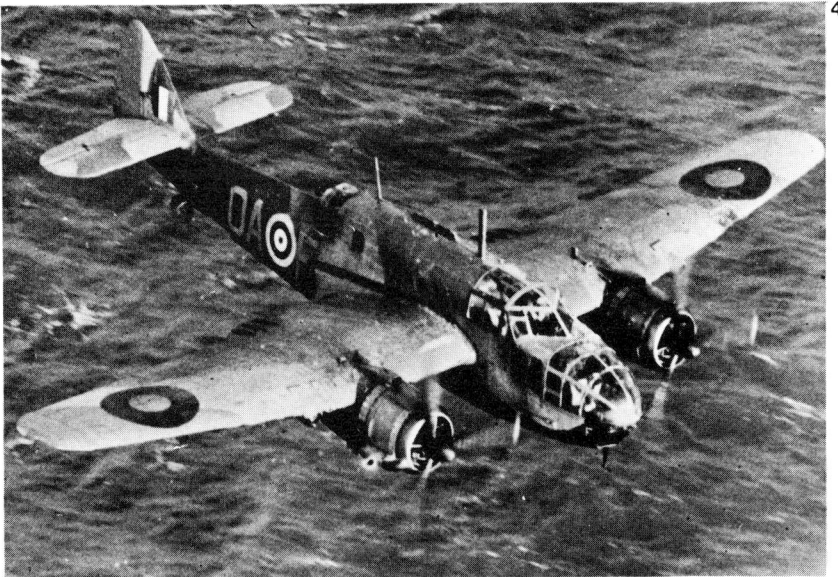

1 Vickers Vildebeest III K4603 served with No 22 Squadron until it crashed in the sea off Bembridge, Isle of Wight on 21 August 1939 while operating from Thorney Island.

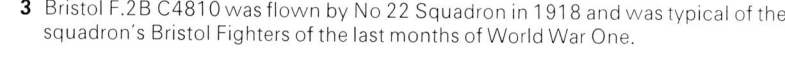

2 F.E.2b Serial No 7703 was one of an early batch built by Boulton & Paul Ltd at Norwich and was a replacement for one of the original batch delivered from the same maker.

3 Bristol F.2B C4810 was flown by No 22 Squadron in 1918 and was typical of the squadron's Bristol Fighters of the last months of World War One.

© Hylton Lacy Publishers Limited

1 Bristol Beaufort W6537 joined No. 22 Squadron on 18 March 1941 but was damaged less than three weeks later and did not return to No. 22 until September. On 30 December 1941, it was sent for repair as a result of action damage and subsequently served with Nos. 86 and 217 Squadrons, being struck off charge in India on 27 July 1944.

2 Bristol Beaufighter X NE604 used in India lacks the extra ventral fin fitted to later Beaufighters of this mark. One of the first Beaufighters to be allotted to No. 22, NE604 was struck off charge on 8 December 1944.

3 Westland Whirlwind HAR.10 XD182 was originally built as a Mark 2 but was later rebuilt to Mark 10 standards. Still in service in 1972, it was the earliest operational Whirlwind at that time.

© Hylton Lacy Publishers Limited

To bridge the time gap between the reporting of enemy ships by reconnaissance aircraft and the launching of a striking force, *Rovers* were initiated. These missions were undertaken by a small number of Beauforts which flew along the enemy-held coast seeking targets. On 15 September, two Beauforts sweeping up the Dutch coast on the first of these missions arrived off IJmuiden soon after dusk. Sergeant Norman Hearn-Phillips (in 'K/22'; L4508) sighted and torpedoed a medium-sized supply ship inside the harbour, the first successful night attack with torpedoes.

Two nights later, six Beauforts were sent to Thorney Island for a night attack on Cherbourg. Becoming separated in cloud, the Beauforts arrived independently over the target to find shipping silhouetted by the flames from dock buildings bombed by a force of Blenheims a few minutes before. Searchlights dazzled the low-flying torpedo-bombers as they dropped their 'tin fish' inside the mole and heavy *flak* came from the ring of forts around the harbour. One Beaufort, the already-mentioned 'K/22', failed to return; but reconnaissance photographs next day showed a torpedoed merchant ship.

Two 'Rovers' on the 18th resulted in a tanker being missed off the Dutch coast while Flight Lieutenant Beauman ran into Bremerhaven in an attempt to hit one of the two 30,000-ton liners—*Europa* and *Bremen*—sheltering there. Blinding searchlights made a drop impossible and it was only with difficulty that the Beaufort was extricated without crashing or hitting an obstruction.

Subsequently 'Rovers' brought occasional sightings and several ships were sunk or damaged. In addition, the squadron diverted some sorties to attack enemy airfields used by night bombers for raids on Britain. The first of these missions, on 15 November, resulted in one of the Cambrai fields in NE France being bombed. Next night, Flight Lieutenant Pat Gibbs (in 'I/22') located Cambrai/Niergnies by its flarepath. This was blacked-out on the approach of the British bomber; but, after the Beaufort's navigation lights were switched-on, it was mistakenly re-illuminated thus obligingly enabling a pair of 250-lb bombs to be placed squarely on the hangars. Beauman (in 'E/22') did similar damage to Vitry-en-Artois and two others bombed Cambrai/Epinoy.

Squadron Leader Roberts (in 'T/22') crashed into the sea during a run-up to a pair of motor vessels off Den Helder on 26 November and the crew was lost. But next day, Flight Lieutenant 'Fanny' Francis ('I/22') hit a 5000-tonner off Arneland and Dick Beauman ('E/22') torpedoed a 7000-ton tanker off the Friesian coast which sank in flames within 10 minutes. Two days later, Gibbs ('S/22'; L9792) hit an 8000-tonner surprised off the Dutch coast. This was one of eight ships in convoy located earlier by two other No. 22 Squadron Beauforts whose attacks had been unsuccessful. The original pair, Beauman and Hicks, rearmed and took off again for another attempt but missed in the face of searchlights and *flak*. Gibbs, after refuelling and collecting another torpedo, led two more Beauforts to the scene of action but failed to find the ships in the darkness. Tired from eight hours of flying, he crashed on landing and both Gibbs and his navigator ended the day in hospital.

A serious loss to the squadron occurred on 5 December when Dick Beauman and his crew (in 'B/22'; L9936) was shot down by *flak* as they made their run into the German naval base at Wilhelmshaven. Four days before the end of the momentous year of 1940, 'X/22' (N1118), carrying 'Fanny' Francis and his crew, failed to return from a 'Rover' over the North Sea.

The new year opened with the same 'Rover' missions, minelaying sorties and occasional attacks on coastal targets. The last-mentioned covered a variety of targets. On 2 March 1941, four Beauforts set out in daylight for the Dutch coast. Of these, 'A/22' bombed Haamstede airfield and IJmuiden jetty, 'E/22' and 'J/22' attacked Borkum airfield, while 'V/22' hit a coaster at Harlingen and suffered a near-miss on a floating crane before going on to Den Oever to bomb barges and a lock gate.

One of No. 22's most successful pilots was Flying Officer Kenneth Campbell. On 13 March, he scored a direct hit with his torpedo on a 3000-ton motor vessel off Borkum, forcing the crew to take to the boats as it sank by the stern. A week later he was not so lucky, being intercepted by a pair of Messerschmitt Bf 110s off Ameland. With his radio operator slightly wounded, 'M/22' had to be crash-landed at base with useless hydraulics.

A second but less fortunate Beaufort from the same mission failed to return.

Campbell was back off the Dutch coast a week later when a 6000-tonner received his torpedo off IJmuiden and he was one of 13 crews who flew down to St Eval at the beginning of April. Across the Channel from Cornwall lay the great naval base of Brest which now harboured a major portion of the German Fleet. The battleships* *Scharnhorst* and *Gneisenau* had both put in after a sweep in the North Atlantic. Fortunately, the convoys which they sighted had each been provided with an escort of an old battleship and escaped the attentions of the notorious 'S and G'. Nevertheless, 22 unescorted merchant ships were sunk during the cruise and, as sighting reports narrowed the search area, the battleships headed for Brest and reached it before the net closed on them.

When a photographic-reconnaissance aircraft confirmed the presence of the German ships on 28 March, a torpedo striking force was detailed to await a further sortie. With the wide expanses of the Atlantic open to them, 'S and G' were a menace to Britain's sea communications, especially as the number of ships available to the Home Fleet which could catch and destroy them was small. The elderly veterans of Jutland sailed individually with the more important convoys leaving only the new battleship HMS *King George V*, the heavily-gunned but slow *Nelson* and *Rodney* and the battle-cruisers *Hood*, *Renown* and *Repulse* capable of engaging the enemy force with the aid of searching cruisers and flying-boats—and the possible intervention of naval or land-based torpedo-bombers to slow the enemy down.

Brest was on the north side of the large Rade de Brest (the 36,000-acre basin) and was protected by coastal *flak* batteries around the anchorage and fighters based at nearby airfields. At night, RAF Bomber Command sent forces of heavy bombers to attack the dockyard in the face of intense AA fire, probing searchlights and dense smoke screens. In the early hours of 5 April, one unexploded bomb landed alongside *Gneisenau* as she lay in drydock. Hurriedly undocked, the battleship was moored to a buoy in mid-harbour while bomb disposal personnel rendered the bomb harmless. At this moment, a Photographic Reconnaissance Unit Spitfire took a photograph of the harbour that was to spell the virtual end of *Gneisenau*'s operational career.

As soon as photographic interpreters located *Gneisenau*'s position, it was clear that there was a possibility of a successful torpedo attack that had not existed before. Within a short time, either the battleship would be back in dock or behind a torpedo net. No. 22 Squadron was within range and was given the task. Wing Commander F. J. St G. Braithwaite doubted whether an attack at night in the face of formidable defences could succeed but was pressed to make the attempt in view of the overriding importance of the target.

Nine Beauforts were available but three were already on a sweep off the Breton coast; and, of these, 'V/22' (N1147) with Hicks and his crew aboard would not return. The remaining six aircraft were to be divided into two waves. The first three aircraft would have bombs to disrupt the torpedo nets while the second wave would carry torpedoes. These last three were manned by Jimmy Hyde from Australia, Ken Campbell from Scotland and Sergeant-Pilot Camp from Ireland, an indication of the mixture of nationalities found in wartime RAF squadrons.

St Eval lacked runways at this time and heavy rain

---

*These 32,000-ton battleships had been announced as '26,000-ton battle-cruisers' to evade treaty obligations.

had swamped the airfield before take-off time. Two bomb-carrying Beauforts became hopelessly bogged-down in the mud and only four managed to become airborne. The Brest Peninsula was covered in mist and low cloud. Consequently, the last of the bomb-laden Beauforts failed completely to find Brest and resorted to dropping its load on a convoy near the Ile de Batz before returning to base with little remaining fuel. Campbell and Hyde arrived over the harbour as dawn broke and awaited the explosion of the bombs which would signal their time of attack. As Hyde circled, another Beaufort passed below on its way into the harbour and was identified as 'X/22', Campbell's aircraft (N1016). Despite the fact that no explosions had been seen as the torpedo nets were, it was hoped, broken open, Campbell was obviously going into his torpedo run.

The chances of success were negligible and there was little point in the second Beaufort following into the now thoroughly-awakened harbour defences. That any surprise was possible at all stemmed not only from the adverse weather conditions, normally precluding such a sneak attack, but also through the fortuitous failure of the bomber Beauforts to arrive ahead to deal with the anti-torpedo nets.

The first sight the Germans had of the torpedo-carrying Beaufort 'X/22' was when it appeared at about 300 feet alongside the mole and diving to 50 feet. This was the signal for a concentrated barrage of *flak* from all quarters; 'X/22' was mortally hit and plunged into the harbour. But not before the crew—Campbell and Sergeants Scott, Mullins and Hillman—had achieved their object. The Beaufort's torpedo ran true and reached its target. Under its explosive impact, the *Gneisenau* lifted bodily in the water.

Then the *flak* opened up again. Sergeant-Pilot Camp

---

1 A Beaufort is started-up at dispersal. The 'trolley-ack' starter unit was standard equipment at all RAF stations to provide power for starting engines. Note the torpedo in its recess under the fuselage and early radar antennae under the wing (*Photo: IWM. CH17131*)

2 A Beaufort sweeps low over a German motorship after dropping its torpedo during an attack on a German convoy off the Netherlands coast (*Photo: IWM. C2635*)

3 Beauforts at North Coates in June 1940. The aircraft running up, L4451, crashed on 11 June 1940 a few days after L9891 in the foreground arrived with No. 22 Squadron (*Photo: IWM. CH646*)

had finally located Brest by flying in at sea-level until the entrance to the *Rade* (basin) hove in sight. Almost immediately, the Beaufort was enveloped in mist and Camp instinctively climbed through the hail of AA fire. The murk thickened and Camp's only chance of attacking had been expended. He returned to base, unaware of Campbell's previous success.

Although not permanently crippled, *Gneisenau* was drydocked at dusk and four nights later was hit again; this time by four bombs dropped during an RAF Bomber Command raid. Almost a year was to pass before, in February 1942, *Gneisenau* eventually emerged from Brest. But then it was only to make the Channel dash to Germany where she was laid-up for the rest of the war.

Meanwhile, *Scharnhorst* remained in Brest; and when the massive *Bismarck* and the heavy cruiser *Prinz Eugen* put to sea in May they were on their own and were hunted down by the Home Fleet. The crews of the two battle-cruisers saw only *Prinz Eugen* enter port after her brief cruise; the biggest battleship in the world was at the bottom of the Atlantic.

The consequences of one torpedo altered the war at sea out of all proportion to the effort expended. A combined force of three modern battleships might have cut the convoy routes to the British Isles and tilted the balance in Germany's favour, especially as *Bismarck*'s sister ship, the *Tirpitz*, was nearing completion. The hard work of the torpedomen had been vindicated in a few minutes of action and Campbell was awarded a well-earned Victoria Cross.

During the time that *Bismarck* was unaccounted for by RAF reconnaissance aircraft, a detachment of No. 22 stood by at Wick in Caithness, Scotland. In June, Beauforts ranged along the length of enemy-held coast attacking shipping; but during the month suffering 20 casualties to its aircrews. One pilot missing (in 'V/22'; X8920) was Sergeant-Pilot Camp. At the end of the month, No. 22 was back at Thorney Island.

Much of the anti-shipping work was undertaken by Hudsons and Blenheims with bombs but No. 22 stuck to the torpedo as its main weapon. Because of the high degree of skill required to make an accurate torpedo attack, bombing was favoured by the Air Staff as being more productive. This did not prevent a number of ships being sunk by No. 22's torpedoes before the end of the year. Wing Commander Braithwaite left the squadron in August to be replaced by Wing Commander J. C. Mayhew who led his first sweep from Leuchars in Fifeshire, Scotland, on 1 September.

Three Beauforts left Leuchars to attack two supply ships and three escort vessels off the Norwegian coast. Two torpedoes were seen to hit the largest vessel which was enveloped in smoke. However, 'W/22' (AW218), with its port engine on fire, crashed in a shallow dive. As well as heavy *flak* coming from the escort, five Messerschmitt Bf 109s arrived on the scene. One ventured too close and caught a long burst of machine-gun fire as it passed slowly by 'V/22'. Turning upside down, it fell away towards the sea as the two remaining Beauforts reached the safety of a layer of cloud.

At the end of October, No. 22 was back at St Eval and bombing Lorient, St Nazaire and La Pallice, all U-boat bases on the Biscay coast. '*Rovers*' were flown along the coast until the end of 1941. On 3 January 1942, instructions were received to prepare to move. The destination was the Middle East.

Going East

At the beginning of 1942, the British Eighth Army and the German *Afrika Korps* faced each other in the sandy wastes of the Western Desert. Both had problems.

Supplies for the British forces in the Middle East travelled vast distances around the Cape of Good Hope to Egypt where they were then trucked into the desert along a tenuous coastal road system. Small quantities of urgent supplies could be airfreighted via Gibraltar and Malta but the bulk of the Eighth Army's ammunition and equipment took months to reach the front line. In contrast, the German and Italian forces were supplied from Italy by a relatively short sea crossing at the end of a good railway system from the enemy's industrial centres. The sea route from Sicily to Tripoli was about 250 miles but a 70 mile crossing between Sicily and Tunisia could bring Italian merchant ships to the relative safety of Tunisian waters under the control of an acquiescent Vichy regime.

Unfortunately for the Axis armies, reality confounded the theorists who saw the Mediterranean as an area dominated by air power and the Italian Fleet. With six modern battleships at the centre of a balanced fleet there was no obvious reason why the Central Mediterranean should not be barred to British ships. With the withdrawal of French North Africa from the war and the occupation of Greece and Crete, there were no bases available to Britain between Gibraltar and the Eastern Mediterranean except Malta. It was an unfortunate exception for the enemy.

Only 50 miles from the coast of Sicily, Malta was the base of a small striking force of naval and RAF aircraft, submarines and, at intervals, cruiser and destroyer squadrons. Attacks by enemy aircraft operating from Sicily failed to neutralize Malta's defensive and offensive potential and every Axis supply ship with stores for Libya had to run the gauntlet of Fairey Swordfish and Albacore torpedo-bombers of the Royal Navy's Fleet Air Arm and Bristol Blenheims and Vickers-Armstrongs Wellingtons of the RAF. After losing three of its battleships in a Swordfish attack on Taranto, one of which never returned to service, the Royal Italian Navy came near to losing its Fleet flagship during the Battle of Cape Matapan in March 1941 when only luck enabled it to escape from a trio of 25-year-old British battleships. By pre-war assessment of air and sea power, the latter should have been seeking refuge in Alexandria harbour, if not out of the Mediterranean completely.

Such reverses caused a complete lack of self-confidence in the Italian Naval Staff and the bulk of Italian naval activity rested on small craft whose ability and dash failed to compensate for the loss of control of the sea lanes by the abdication of the main fleet. German dive-bombers made operations by the Mediterranean Fleet difficult but failed to subdue Malta or stop attacks on the Axis seaborne supply lines to Libya.

It was into this scene that No. 22 Squadron was to be fitted. Japan had entered the war in December 1941 and it was planned that No. 22 should go to India. But the need for torpedo-bombers in the Mediterranean resulted in the squadron being diverted. Since the outbreak of war in 1939, No. 22 had claimed the destruction of 100,000 tons of enemy shipping for the loss of 154 killed in action, missing or killed in accidents with 12 more known to have been captured and made prisoners-of-war. Its experience in anti-shipping operations would be invaluable in the Mediterranean.

On 7 January 1942, 476 ground personnel left for Liverpool. All aircrew were attached to No. 86 Squadron to carry out fuel duration tests on their Beauforts in preparation for the long flights ahead. The seaborne party sailed on the transport *Ormonde* on 16 February and arrived at Suez on 16 April. It was 18 March before the first flight of seven Beauforts took off from Portreath in Cornwall and headed out to sea for Gibraltar. Six arrived safely but the seventh was posted missing. Next day, the six aircraft left for Malta but were intercepted by Bf 110s near the island. Pilot Officer White's Beaufort was damaged and the rear gunner wounded, a crash-landing being made at Luqa. After staying overnight, two aircraft flew on to Egypt, followed by one on each of the succeeding days. On the 23rd, White's Beaufort also reached Egypt. Other aircraft completed the long journey from England; and, by the end of the month, the last had arrived from the first flight while the remainder trickled in one by one early in April.

On 4 April, No. 22 despatched its first strike from the advanced airfield at Sidi Barrani. Six aircraft were sent to intercept a convoy which had been spotted by reconnaissance aircraft and one ship was hit. On 17 April, Lander led a mixed force of aircraft from Nos. 22

and 39 Squadrons to a convoy reported by another 22 Squadron crew. This latter aircraft, captained by Flight Sergeant Howroyd, made for Malta after passing on the position of the enemy ships but was followed in by a Bf 109. Riddled with cannon fire, the Beaufort landed and rolled to a halt—with the pilot and navigator dead.

The eight Beauforts of the striking force had an escort of four Bristol Beaufighters. Not locating the convoy immediately, they began a square search of the area. Soon, a vanguard of two Junkers Ju 88s escorting the convoy appeared but failed to see the escorting Beaufighters which immediately shot them down. As a result, they were now low on fuel and accordingly were forced to signal their departure to the Beauforts. Soon after, the small convoy was sighted and the torpedo-bombers headed towards the four Axis merchant ships. Overhead, the crews could see an air 'umbrella', about twenty escorting fighters.

Disregarding the odds and flying through a curtain of *flak*, the Beauforts released their torpedoes and, under ceaseless attack from the escorting fighters, made for the safety of Malta. Five were lost before the island was reached and of the trio of surviving Beauforts, only one could be repaired to fly again. Though none of the crews knew it, this was to be the last time No. 22 used its torpedoes against an enemy ship.

Within days, the squadron was on the move once more. Eight aircraft left for Habbaniya on the journey to Shaibah, Sharjah, Karachi, Hyderabad and Ratmalana in Ceylon. The flight began on 21 April and on 28 April one solitary Beaufort arrived in Ceylon, leaving a trail of ailing aircraft along the way. Six more trickled in during May and by the end of the month the seaborne party had arrived.

Ceylon was a backwater but nobody was aware of the fact at the time. In April 1942, a Japanese carrier force had made two devastating raids on Colombo and the naval base at Trincomalee. The small force of Blenheims on the island had failed to damage the enemy but had torpedo-bombers been available, they might have had more effect. Sent to fill this gap, the Beauforts awaited a further attack but none came and the months were filled with training exercises in co-operation with the Eastern Fleet and with the fighter squadrons.

The end of an era came in May 1944. On 12 May, the squadron's first Bristol Beaufighter flew into Ratmalana. More arrived in the following weeks and on 23 June the squadron's complement was complete. The last of the Beauforts were flown away for breaking-up, leaving behind a tradition of service that few squadrons have equalled.

Beaufighters almost turned the wheel full cycle. Once more No. 22 Squadron had two-seat fighters but they were very different in concept from the early 'Fees' and 'Biffs'. Originally designed as a long-range fighter for night-fighting and escort duties, the Beaufighter was at first powered by two Bristol Hercules air-cooled, two-row radials of 1400 h.p. each. The pilot sat in the nose behind an armoured windscreen and had an armament of four 20-mm Hispano cannon and six 0·303-in Browning machine-guns. Behind him sat the navigator/radar operator under a small Perspex blister.

RAF Fighter Command had first call on Beaufighter production to equip night-fighter squadrons but some were released to RAF Coastal Command in April 1941 and these began to replace Blenheims for long-range escort missions and defensive patrols over shipping. By the end of the year, it had been decided to replace Beauforts by fitting torpedo gear to Beaufighters and by the end of 1942 enough had been supplied to equip No. 254 Squadron. On 4 April 1943, torpedo-fighter (TF) 'Torbeaus' sank their first ships.

The version received by No. 22 was the TF. Mk X. Powered by Hercules XVII radials of 1735 h.p. each, the heavily-laden TF. Mk X had a maximum speed of over 300 m.p.h. The six machine-guns in the wings had been removed but a single 0·303-in Vickers K-gun was provided for the observer. As well as a torpedo, there

---

1 The naval dockyard at Brest photographed by a Spitfire of the RAF Photographic Reconnaissance Unit. One battlecruiser is in drydock (No. 1 right) and the other is alongside the fitting-out quay (No. 1 left). A torpedo net is in place (3) and it was the absence of this that allowed a No. 22 Squadron Beaufort to torpedo *Gneisenau*. Bombs have damaged dockyard buildings nearby (2) and the base fuel tanks (4) (*Photo: IWM. C2086*)

2 Additional defensive armament was fitted to Beauforts for rearward defence. A pair of beam guns was installed just forward of the dorsal turret which has the original single-gun arrangement. The black rectangles are access toeholds for reaching the upper hatch and were usually covered by spring-loaded flaps (*Photo: IWM. CH637*)

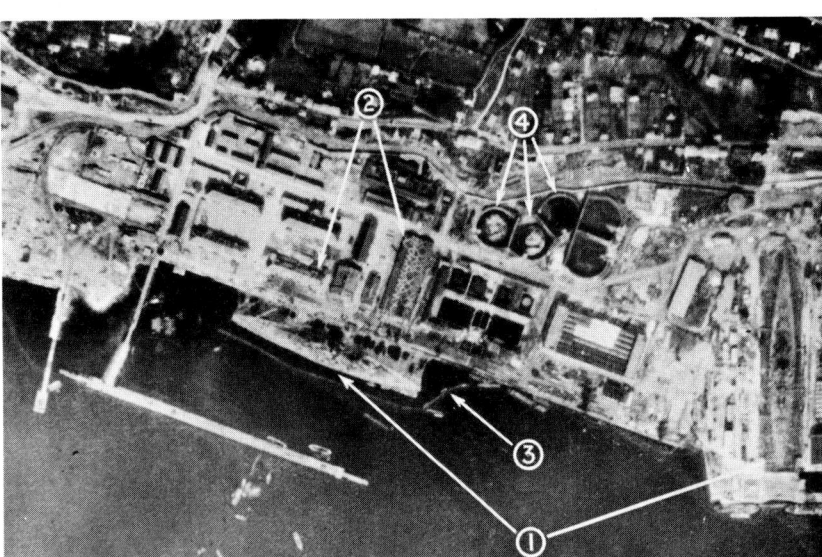

was provision for alternative loads of rockets or bombs. Coastal Command's strike wings operated a mixture of variants; cannon and rocket-armed Beaufighters were used to supress *flak* and attack small craft while bomb-laden and torpedo-equipped aircraft dealt with larger ships. The same formula was adopted for the Ceylon-based strike squadrons but by the time they were operational, the torpedo was out of favour and rocket projectiles were the Beaufighter's main offensive weapon.

During August 1944, No. 22 ceased to be a torpedo-bomber squadron. Until the crews could undertake live rocket-firing practice—and supplies of rockets were short—the squadron flew coastal patrols and covered the return of US Army Air Forces Boeing B-29 Superfortresses from their raid on the oil refineries at Palembang in Sumatra.

In December 1944, came a promise of action in the near future. On 23 December, Beaufighters led by Wing Commander J. M. Lander left Ceylon for Kumbhirgram on the Burma front. Of the 11 aircraft, one (NE407) suffered from engine failure over the Kistna Delta and it was two days before a Catalina located and rescued the occupants. Another (NV328) had an engine cut on take-off from Vavuniya and ended upside-down at the end of the runway.

On 27 December, Wing Commander Lander and Sergeant Berry took off for a reconnaissance over the Imphal area to resume active operations. The squadron's ground echelon left next day for the traditional slow journey up the east coast of India. Accommodation on the train was primitive and it suffered from the normal hazards of train travel in India. The engine ran out of water and it took four hours to locate a supply. The train crew decided that their shift had ended while still some 20 miles from a station; but were effectively persuaded otherwise. At Madras, fortunately, an efficient traffic officer organised the squadron into overnight camp in time to celebrate the New Year. Next day they were back on the railway for more faltering journeys. Thirteen days later, the exhausted airmen reached their destination. Their new base was a wide grassy airstrip 1800 yards long. Accommodation was in wooden *bashas* well dispersed around the airfield and there was a shortage of water.

As part of No. 901 Wing, the squadron was given the task of attacking Japanese communications and camps. This entailed on most occasions strafing river craft as road and rail bridges were the constant targets of day bomber squadrons. On 18 January 1945, six Beaufighters set off on No. 22's first *Rhubarb*, the codename for this type of operation.

Two days later, the squadron suffered an unexpected loss. Wing Commander Lander, who had first joined No. 22 as a Pilot Officer in May 1941, was killed when his jeep fell into a stream near the airfield. Apart from a six-month rest he served with the squadron continuously and was awarded the Distinguished Flying Cross for his leadership during the attack on a Mediterranean convoy in April 1942. Squadron Leader R. Gee took his place from being flight commander of 'A' Flight. On the same day, a Beaufighter ('J/22') was posted missing.

Low-level attacks were inherently risky. Apart from return fire from the ground, skill in gauging an aircraft's altitude during the firing run was necessary to avoid trees or hitting the glassy waters of Burmese rivers and lakes. On 3 February, 'J/22' (a replacement 'J') was hit by ground fire near Magwe and crashed into the riverbank in flames. 'T/22' developed engine trouble and ditched near Cox's Bazaar. The crew were never found. On the 10th 'Q/22' failed to return from patrol over the Gulf of Martaban and, on the 19th 'T/22' (replacement 'T') flew into the water, probably deceived by haze and a glassy calm.

In March, the squadron flew 138 sorties, losing two aircraft; but Squadron Leader Fenton and Flight Sergeant Kane (from 'R/22') were rescued by the army near Pegu. Two seagoing ships had been destroyed during the month and eight more were damaged. Over 700 small craft had been strafed. During April, two more crews were lost during 104 missions. 'P/22' failed to return from a *Rhubarb* while the crew of 'W/22' were captured and killed by the Japanese. Their Beaufighter had been inadvertently set on fire by a USAAF North American P-51 Mustang while attacking river craft near Henzada.

After transferring to No. 346 Wing on 18 April, the squadron saw the beginning of the end for the Japanese in Burma when 17 Beaufighters of Nos 22, 177 and 211 Squadrons, armed with rocket projectiles, attacked gun positions defending the mouth of Rangoon River. They repeated this action on the following day as landing craft took an invasion force to capture Rangoon. On 4 May, the squadron suffered its last operational loss when 'C/22' did not return to base. A final patrol was flown on 15 May 1945; but there were no targets left within range and operational flying was suspended.

On 12 June 1945, the main party of ground personnel left for Chittagong and on 21 June, 15 Beaufighters and a North American Harvard advanced trainer (and squadron 'hack') flew to Gannavaram, near Madras, where they were joined by No. 217 Squadron to form No. 907 Wing. News of the Japanese surrender meant that the wing would no longer be required for the invasion of Malaya and the welcome intelligence was duly celebrated. It came as a shock, however, when the Commanding Officer received a letter on 8 September

A Whirlwind HAR.10 rescue helicopter demonstrates its winch fitted to the starboard side. As the area under the aircraft is invisible to the pilot, a crew member is stationed at the door to relay information to the cockpit (*Photo*: Air-Britain Digest)

advising him that the squadron would be disbanded. Within a fortnight, Beaufighters began to leave for No. 308 Maintenance Unit at Allahabad and personnel were posted. By the end of the month, another famous squadron had been disbanded and the '*Dinky-Dos*' ended another chapter of their eventful history.

But the Far East had not seen the end of No. 22. On 1 May 1946, the remnant of No. 89 Squadron at Seletar was renumbered to revive the famous '22'. There were no aircraft and No. 89 had been reduced to flying Seletar's Supermarine Walrus amphibians on occasional air-sea rescue flights. Therefore, it was with some relief that instructions were received to send seven aircrew—with three from No. 84 Squadron—to Karachi to pick up five de Havilland Mosquito VI fighter-bombers as an interim measure pending the receipt of night fighters. Four had arrived by mid-June and flying was resumed.

There were now, fortunately, few incidents of note as in peacetime these were usually disasters of varying magnitude. However, Flying Officer McLachlan created a certain stir when he went on leave in the Cameron Highlands, north of Kuala Lumpur, Malaya. Going for a walk 'to work up an appetite for lunch', he succeeded beyond his wildest dreams. Hopelessly lost, he was found a week later sitting on a log in the jungle, having existed for all this time on wild fruit and herbs. He was taken to hospital suffering, not unnaturally, from exhaustion. Nobody in No. 22 went for walks in the jungle thereafter.

On 15 August 1946, No. 22 disbanded again, once more a victim of changing post-war policies.

Flying Lifeboats

In April 1953, the Royal Air Force acquired its first helicopter search and rescue squadron when No. 275 Squadron was formed with Bristol Sycamores. Although helicopters had been used since before the end of World War Two, post-war policies in this field were hampered by shortage of equipment. Although successful helicopters were in prospect at the beginning of the war, all efforts were soon to be concentrated on combat types and the promising G. & J. Weir*—of Scottish origin—designs were dropped. In the United States, development continued and the Sikorsky experimental helicopters were transformed in a short time to a state where they could be used operationally, though under many restrictions. Some USAAF Sikorsky R-4Bs and R-6s were transferred to the Royal Air Force and Royal Navy under Lend-Lease and were flown experimentally under the name 'Hoverfly' for some years. British development was resumed and the first to reach production status was the Bristol Type 171 Sycamore.

As the first of the line, the Sycamore suffered from its share of teething troubles. Westland Aircraft chose an alternative course by obtaining a licence to build Sikorsky S-55 helicopters at Yeovil, Somerset. Using the basic airframe, rotor assembly and engine, Westlands incorporated British equipment to meet RAF and Royal Navy standards and began production under the name Whirlwind. Their investment proved to be a shrewd one. Today, Westland Helicopters is the sole major manufacturer in Britain, having taken over the helicopter divisions of both Bristol and Saunders-Roe.

On 15 February 1955 No. 22 Squadron was reformed as the first RAF Whirlwind squadron. Its base was to be a return, after nearly 14 years, to Thorney Island; where the Air-Sea Warfare Development Unit was experimenting with helicopters for anti-submarine use. As no Whirlwinds were yet available, some training was put in on a Sycamore (WV 781) of that unit until Whirlwind HAR.2s began to arrive in June 1955. Initially, only a Headquarters and two flights were formed and on receipt of its Whirlwinds, 'B' Flight moved to Martlesham Heath in Suffolk. At the end of September, 'C' Flight was formed at Valley in Anglesey.

The squadron's primary duty was to rescue military aircrew from both the sea and from inaccessible places which ground transport could not reach quickly. An additional task, which soon overshadowed the primary one in scope, was the rescue of anyone in distress on the sea or lost or injured in rugged terrain. To carry out rescues, Whirlwinds needed a crew of three; pilot, winch-operator and winchman. Unfortunately, from his position above the nose, the pilot could not see directly below the helicopter and so the winch-operator was required to maintain a running commentary to enable the pilot to hover in the correct position at the right height. The winchman was lowered on a line to the survivors and could attach a bridle to each and have them hoisted aboard the aircraft. As the winchman's position was exposed, especially at sea and in wintry conditions, the winch-operator was also trained to carry out the same tasks in case of injury or fatigue.

On 4 June 1956, squadron headquarters moved to St Mawgan in Cornwall. In consequence, 'D' Flight was formed to remain at Thorney Island when 'A' Flight departed for St Mawgan. 'B' Flight was transferred about the same time to the old flying-boat station at Felixstowe where RAF Air-Sea Rescue craft were based. When this RAF Station was closed in 1961, 'B' Flight moved to Tangmere in Sussex; and, three years later, back to Thorney Island again—'D' Flight having closed down there at the end of 1959. 'A' Flight had moved up the coast to Chivenor, Devon, in November 1958 as the operations of No. 229 Operational Conversion Unit were more intensive than those of RAF Coastal Command at St Mawgan and nearby St Eval. The Hawker Hunter jet fighters on training sorties were also more likely to need rapid rescue than Coastal's aircraft.

Rescue duties in crowded South-East England became the responsibility of 'D' Flight when it was re-established on 1 July 1961 at Manston in Kent, where it remained until March 1969. It was then withdrawn to provide Whirlwinds for a search and rescue unit at El Adem, Libya. This caused an outcry by many local authorities in the region who found their popular holiday beaches left without the comforting presence of rescue helicopters. The Ministry of Defence's reply was that the search and rescue helicopters were intended for military use and military flying in the area was negligible.

In 1962, No. 22 re-equipped with improved Whirlwinds. The Pratt and Whitney R-1340 piston engines on earlier Whirlwinds had been replaced by Bristol Siddeley (DH) Gnome turboshaft units which gave the type a new lease of life. Many Whirlwinds were re-engined and crews found them much smoother aircraft after modification.

Survival missions were numerous and during the first 10 years of search and rescue by RAF helicopters about 1100 people, the majority civilians, were rescued in the course of over 3000 operational sorties.

Rescues continued at this rate around the coasts and from the thousands of rescues, one may be chosen to illustrate the problems faced frequently by Whirlwind crews. One Sunday afternoon in May 1970, a swimmer got into difficulties off the North Devon coast. The Chivenor 'A' Flight stand-by helicopter flew to the scene, successfully winched-up the swimmer and set down in the grounds of Barnstaple hospital. As they handed him over to the hospital staff, another call sent the Whirlwind across the Bristol Channel to Pwlldu Head where a man had fallen onto a narrow ledge 80ft above the sea and was seriously injured. The only way to rescue him was to hover with the rotor blades only inches from the cliff face and to lower the winchman with a stretcher from 60ft above the ledge in turbulent air conditions. Fuel was jettisoned to increase the available power and the port cabin window unshipped to enable the winch operator to lean out and judge the rotor clearance. It took 25 minutes of tension to get the injured man off his ledge and fly him to hospital. The Whirlwind crew then headed back to Chivenor.

Thirty minutes after arrival, the same crew were back on the alert as a cliff rescue team and an inshore rescue boat headed for a youth injured while swimming from rocks. Neither could reach him and the indefatigable Whirlwind was once more in the air. In failing light and severe turbulence, the pilot had to hover out of wind between a 300-ft cliff and a rock pinnacle, using the full length of the cable to hoist the injured person aboard. After passing him on to the Coastguards on the cliff-top, the Whirlwind crew once more made for base only to be recalled to the scene where two members of the cliff rescue team were now stranded on the cliff. It was now completely dark and the helicopter's searchlight and landing light had to be used to locate the position of the men who were also in the dangerous chasm between cliff and pinnacle. Nevertheless, both were successfully uplifted and taken to the cliff-top. This time the exhausted crew succeeded in getting back to Chivenor for a well-earned rest. Subsequently both pilot and winch-operator were awarded a Queen's Commendation.

Since it formed at Thorney Island in February 1955, No. 22 Squadron has answered over 5000 calls for assistance and has saved nearly 2700 lives. The squadron's detached flights maintain a 24-hour stand-by; 15-minute notice during the day and 1-hour notice after nightfall. Perhaps the unofficial motto of 1934 (*Semper resurgam*—'I shall always rise again') most aptly describes the contemporary No. 22 Squadron—ever ready to answer calls for succour from those in grave peril both at sea and on land.

---

*G. & J. Weir Ltd of Glasgow, was awarded an October 1938 design and one prototype (serial R5269) contract by the Air Ministry for a two-seat, two-rotor (on outrigger booms) helicopter powered by a 200 h.p. DH Gipsy Six inline. First flight, 27 October 1939. In July 1940, after some 70 hours of flight testing, the development programme was shelved because of the war situation.—EDITOR.

SQUADRON BASES

Gosport, Hampshire	1 September 1915
St Omer, France	1 April 1916
Vert Galand, France	1 April 1916
Bertangles, France	16 April 1916
Chipilly, France	27 January 1917
Flez, France	1 May 1917 (two Flights only)
	1 July 1917 (complete)
Warloy, France	3 July 1917
Izel-le-Hameau, France	5 July 1917
Boisdinghem, France	14 August 1917
Liettres, France	10 September 1917
Auchel, France	22 January 1918
Treizennes, France	2 February 1918
Serny, France	21 March 1918
Vert Galand, France	23 March 1918
Serny, France	10 April 1918
Maisoncelle, France	30 July 1918
Izel-le-Hameau, France	22 October 1918
Aniche, France	26 October 1918
Aulnoy, France	17 November 1918
Witheries, France	22 November 1918
Nivelles, Belgium	20 December 1918
Spich, Germany	21 May 1919
Ford, Sussex	1 September 1919 to
	31 December 1919
Martlesham Heath, Suffolk	24 July 1923 to
	1 May 1934
Donibristle, Fifeshire	1 May 1934
Hal Far, Malta	10 October 1935
Donibristle, Fifeshire	29 August 1936
Thorney Island, Sussex	10 March 1938
North Coates, Lincolnshire	8 April 1940
Thorney Island, Sussex	25 June 1941
St Eval, Cornwall	28 October 1941
Sea party departed for Middle East	16 February 1942
Aircraft departed for Middle East	18 March 1942
Fayoum Road, Egypt	20 March 1942
Ratmalana, Ceylon	28 April 1942
Minneriya, Ceylon	30 September 1942
Vavuniya, Ceylon	15 February 1943
Ratmalana, Ceylon	21 April 1944
Vavuniya, Ceylon	7 July 1944
Kumbhirgram, India	23 December 1944
Joari, India	26 January 1945
Chiringa, India	18 April 1945
Gannavaram, India	21 June 1945 to
	30 September 1945
Seletar, Singapore	1 May 1946 to
	15 August 1946
Thorney Island, Sussex	15 February 1955
St Mawgan, Cornwall	4 June 1956

Detached flights as follows:
'A' Flight:
Thorney Island, Sussex	15 February 1955
St Mawgan, Cornwall	4 June 1956
Chivenor, Devonshire	4 November 1958 to date

'B' Flight:
Thorney Island, Sussex	15 February 1955
Martlesham Heath, Suffolk	25 June 1955
Felixstowe, Suffolk	16 May 1956
Tangmere, Sussex	1 June 1961
Thorney Island, Sussex	5 May 1964 to date

'C' Flight:
Valley, Anglesey	27 September 1955 to date

'D' Flight:
Thorney Island, Sussex	4 June 1956 to
	31 December 1959
Manston, Kent	1 July 1961 to
	31 March 1969

SQUADRON EQUIPMENT
Period of Use and Typical Serial and Code Letters

Blériot	September 1915 to March 1916	4654
B.E.2c	September 1915 to March 1916	2053
Maurice Farman Shorthorn	September 1915 to October 1915	2944
Curtiss J.N.3	October 1915 to January 1916	6117
B.E.8A	October 1915 to November 1915	2133
Caudron G.III	October 1915	5258
Martinsyde S.1A	October 1915 to March 1916	5442
Bristol Scout	December 1915 to March 1916	5294
Avro 504A	January 1916 to March 1916	2890
F.E.2b	February 1916 to August 1917	5224
Bristol F.2B	July 1917 to August 1919	B1152 (C)
Vickers Vildebeest I	May 1934 to October 1935; September 1939 to November 1939	K2810
Vickers Vildebeest III	May 1935 to February 1940	K4187
Vickers Vildebeest IV	March 1938 to February 1940	K6396
Bristol Beaufort I, II	November 1939 to June 1944	N1081 (OA-U)
Bristol Beaufighter X	May 1944 to September 1945	NE604 (OA-A)
de Havilland Mosquito FB.6	May 1946 to August 1946	
Westland Whirlwind HAR.2	June 1955 to August 1962	XJ725
Westland Whirlwind HAR.10	August 1962 to date	XP352

COMMANDING OFFICERS

Captain The Hon Lord Lucas	1 September 1915
Major R. B. Martyn	22 September 1915
Major L. W. Learmount MC, DSO	26 January 1917
Major J. A. McKelvie	11 March 1918
Major V. A. H. Robson	10 February 1919
Major J. C. Quinnell	23 July 1919 to
	31 December 1919
S/Ldr C. H. Nicholas DFC, AFC	24 July 1923
S/Ldr T. H. England DFC, AFC	2 March 1925
S/Ldr E. R. L. Corballis OBE, DSO	3 March 1927
S/Ldr J. Noakes AFC, MM	1 August 1927
S/Ldr E. S. Goodwin	8 January 1929
S/Ldr C. E. Maitland DFC, AFC	20 March 1933 to
	1 May 1934
S/Ldr T. A. Warne-Brown	1 May 1934
S/Ldr R. J. M. St Leger	12 September 1935
S/Ldr W. G. Campbell	27 April 1937
S/Ldr M. V. Ridgeway	2 April 1938
W/Cdr H. V. Mellor MVO	28 November 1939
S/Ldr R. E. X. Mack DFC	26 May 1940
W/Cdr F. J. St G. Braithwaite	3 June 1940
W/Cdr J. C. Mayhew	31 August 1941
W/Cdr W. A. L. Davies	2 November 1941
W/Cdr J. Bateson	7 January 1942
W/Cdr J. M. Lander DFC	11 September 1943
S/Ldr R. Gee	20 January 1945 to
	20 September 1945
S/Ldr A. G. A. Good	1 May 1946 to
	15 August 1946
S/Ldr P. C. Bowry	15 February 1955
S/Ldr J. R. Ritchie AFC	25 January 1957
S/Ldr A. P. Dunn	18 November 1958
S/Ldr G. L. Verran	17 March 1961
S/Ldr L. A. Barber	15 June 1962
S/Ldr K. Annable	20 July 1964
S/Ldr D. A. W. Todman DFC	5 August 1966
S/Ldr A. Salter	15 July 1968
S/Ldr J. Weaver	3 August 1970

Beaufighter TF.X torpedo-fighters were fitted with an extra gun in the observer's cockpit for rear defence. Later version of the Mark X had dorsal fillets to the fin to reduce tail swing on take-off. (*Photo: IWM. CH18583*)

201 Squadron

A formation of Southamptons off Shanklin, Isle of Wight. The leading aircraft is K2965, a Mark II; the last Southampton built for the RAF
(*Photo:* Flight International *19022*)

Change of Blue

Ever since the German armies had rolled over the Belgian plain in 1914, the French port of Dunkerque had been the home of units of the Royal Naval Air Service. From a handful of miscellaneous aircraft, the force had expanded to comprise squadrons of fighters, day bombers, night bombers and seaplanes. Based at airfields like Bray Dunes, Ste Marie Cappel, Teteghem, Petite Synthe, Furnes and others near the coast, the RNAS squadrons served the British and Belgian troops hanging on to the last piece of Belgian soil around Ypres which had not been occupied by the enemy.

Among them was No. 1 Squadron, RNAS. At first, it was engaged in the normal round of reconnaissance and patrolling with a mixture of aircraft types but, in June 1916, it became the first squadron to be equipped with Sopwith Triplanes. This type came into service at a time when 'Number One Naval' had lost its identity in a larger formation known as No. 1 Wing, RNAS— which had come into being a year before during a reorganization of the naval units. Two flights of this wing were known as 'A' Squadron and in July 1916, this became known as 'Detached Squadron'. Operating as an independent unit, the squadron became more and more involved in purely military flying, being much-needed support for the hard-pressed Royal Flying Corps squadrons beside which it flew. The three sets of mainplanes of the Triplane belied its powers of climb and manoeuvre and the squadron began to establish its reputation as one of the most successful fighter units on the Western Front. On 6 December 1916, it recovered its original number, become No. 1 Squadron, RNAS or 'Naval One', to prevent confusion with No. 1 Squadron, RFC, operating nearby.

Despite its climbing power, the triplane was replaced by the Sopwith Camel in December 1917. The squadron's new biplanes may have been nore manoeuvrable than 'Naval One's' former mounts but there was no doubt which one the pilots preferred. By comparison with the docile Triplane, Camels had vicious tendencies, notably

27

in turning. Due to torque from the rotary engine, turning in one direction was as rapid as it was sluggish in the opposite direction. Inexperienced pilots often found themselves in a spin from which many had insufficient training to recover. Nevertheless, the Camel ended the war with the highest number of enemy aircraft destroyed to its credit than any other type.

The days of the Royal Naval Air Service were numbered. On 1 April 1918—which many considered an appropriate date—both the RFC and RNAS were to lose their identities in a new service, the Royal Air Force. It was not a move which generated much enthusiasm among naval personnel. They were part of a service with hundreds of years of tradition behind it, unlike a large part of the RFC which had itself only existed for six years. Many of its officers had been seconded from regiments in which they had served for only short periods in a wartime environment. To them, the RFC was just another part of the army and the number of pre-war officers surviving in squadrons was small.

On the other hand, the naval pilots were representatives of a vast fleet and stood out among the military uniforms seen everywhere in France. With the end of the war, the RAF would, doubtless, be disbanded in the same way as other specialized units such as the tank squadrons, salvage corps and concert parties. This would leave naval officers transferred to the RAF either out of a job or returned to the Navy with a gap in their careers. However, trained pilots were of more use in the cockpits of their aircraft than on the bridges of ships, and their Lordships of the Admiralty were not too interested in aircraft based in France in any case. So the RNAS was transferred entirely to the new service. Little did its original members know what history had in store for their new force.

Because existing RFC squadron numbers had started at No. 1, the RNAS squadron numbers duplicated these. In consequence, it was decided that all former RNAS squadron numbers should have 200 added on to them; thus, on 1 April 1918, No. 1 (Naval) became No. 201 Squadron, Royal Air Force.

At the time, few units had time for any ceremony. The German offensive in March had torn huge gaps in the Allied line and the enemy had flooded through almost to Amiens. Every RAF squadron was engaged in trying to hinder the enemy advance by machine-gunning and bombing columns of troops, artillery and supply vehicles. No. 201's Camels could carry small bombs and took part in ground attack missions in addition to maintaining patrols over the battle area to keep German fighters away from the low-level units.

During May, 112 combat reports were submitted by pilots of the squadron, recording the destruction of eight enemy aircraft and 16 more being seen to go down apparently not under control. In its ground-hugging sorties, Camels carried four 25-lb. Cooper bombs and took it in turns to patrol over a given area to keep the enemy under constant attack. The wooden and unarmoured Camels were subjected to considerable ground fire. Many were written-off, even after returning their pilots to their home airfield, because of multiple hits from small arms fire.

On August 12, No. 201 lost its Commanding Officer, Major C. D. Booker, when he was shot down over the front line. Major C. M. Leman took over the squadron for the remaining months of the war. At the same time, the Allies took the offensive around Amiens and the German army began to retreat for the last time. Ground-attack missions continued, this time to hamper the enemy retirement.

During October, Major W. G. Barker was attached to No. 201 with his Sopwith Snipe. The Snipe was an improvement on the Camel which was just coming into service and was eagerly awaited by the Camel squadrons who wanted a scout as handy as the Camel but without its unforgiving nature. On 27 October, Barker was on patrol (in E8102) when he sighted and attacked a German observation aircraft over the Forêt de Mormal,

near Ypres. After a short burst from the Snipe's twin Vickers guns, it broke up in the air. Simultaneously, Barker was hit in the right thigh by a bullet from a Fokker D VII that had dived on him unseen. Despite his wound, Barker succeeded in out-turning the enemy fighter and shot it down in flames.

Before he could make for home, Barker's Snipe was in the middle of a large formation of Fokkers with one object in mind. Fortunately, the very number involved, estimated as about 20, resulted in the enemy pilots getting in each others' way. The lone Snipe pilot had no such problem and succeeded in sending two Fokkers down in spins despite another wound in the left thigh. Losing consciousness, Barker spun down, recovering to find himself once more among a large formation of German fighters. Selecting one, he fired off his remaining ammunition and saw it burst into flames before diving for home. Yet another flight of fighters tried to cut him off but he finally evaded them and reached the British lines where he crashed the tattered remains of his faithful Snipe. For this exploit, Barker was awarded the Victoria Cross, the second awarded to the squadron. The first had been earned by Flight Sub-Lieutenant R. A. J. Warneford in June 1915 for the destruction of the German airship LZ37 near Bruges.

1 Major Barker's Snipe after being recovered from the Front Line after the adventures which earned him the Victoria Cross. The only Snipe to serve with No. 201, E8102 carried a distinctive individual marking   (*Photo: IWM. Q69644*)

2 A Southampton (S1249) takes-off from Calshot with an extended radio mast. The numerals of the aircraft serial number were carried large on the bows as identification and the original squadron badge is painted under the nose gunner's cockpit

3 Five Southamptons of No. 201 Squadron fly past the Landslip, Isle of Wight during a formation flight around the island   (*Photo: Flight Int'l. 10178S*)

4 A Southampton (S1301) heads up Spithead for Calshot. Well displayed are the staggered rear gunners' cockpits just forward of the roundel and the uncowled Napier Lions
(*Photo: Ministry of Defence H327*)

The Armistice came on 11 November 1918 and all operations ceased. On 23 January 1919, No. 201 was reduced to a cadre unit and handed over its aircraft to No. 203 Squadron. Next month, its remaining personnel returned to England and were stationed at Lake Down, near Salisbury, Wiltshire, until disbanded at the end of the year.

The Big Boats

For ten years of peace, the seaplane station at Calshot had been the home of most of the RAF's home-based flying-boats. Situated on a spit at the entrance of Southampton Water, Calshot consisted of a cluster of hangars around the Martello Tower that provided not only accommodation for the headquarters staff but also a vantage point to watch the stretch of water used by the seaplanes and flying-boats. Lacking sufficient area for living accommodation, barracks and messes were built on the mainland and a narrow-gauge railway transported personnel and stores between the two parts.

The quickest method of transport to Southampton was to signal the Isle of Wight ferry to slow down and transfer passengers from a seaplane tender.

Calshot was the home of the School of Naval Co-operation and Aerial Navigation at the end of the war and retained this role, with the Seaplane Training Squadron training marine airmen (in co-operation with Lee-on-Solent further down Spithead) and the Air Pilotage Flight. Its operational element was No. 480 (Coastal Reconnaissance) Flight and this unit was renumbered No. 201 Squadron on 1 January 1929.

Standard equipment for RAF flying-boat units at this time was the Supermarine Southampton. Powered by two 500 h.p. Napier Lion engines, the 75-foot wing-span biplane had a crew of five housed in separate cockpits in the Mark II's metal hull. With a maximum speed of just over 100 m.p.h., the Southampton had a range of 770 miles with normal fuel. Armament consisted of three 0·303-in Lewis guns, one in the nose and two in staggered dorsal gun positions; it could also carry 1,100 pounds of bombs.

No. 201's aircraft had already proved their worth. At the end of 1927, four Southamptons of the Far East Flight had left for a 27,000-mile cruise to the Far East, visiting India, Australia and Hong Kong before settling at Singapore to become No. 205 Squadron. For the first few years, the squadron took part in exercises with the Home Fleet and experimented with various pieces of equipment of varying degrees of utility intended to aid the flying-boat crews.

In July 1929, No. 201 appeared in public for the first time when it flew over the annual RAF Hendon Air Display, having prudently reconnoitered the Welsh Harp nearby just in case it was required as an emergency haven. At the end of the month, the squadron left on the first of its cruises that became such a feature of life in flying-boat squadrons during the 1930s. First stop was at Donibristle, on the Firth of Forth, where over three weeks were spent at the armament training camp at Leuchars practising air gunnery and bombing on the Tentsmuir ranges.

The next staging was to Invergordon on Cromarty Firth, one of the Royal Navy's fleet anchorages, followed by a flight to Stranraer; also destined to be a wartime flying-boat base. The return flight via Pembroke Dock was disorganized by fog but three boats put down at Holyhead on Anglesey and one at Newquay, Cardiganshire. Next day they proceeded to Pembroke Dock where the small naval establishment was being opened up as a new flying-boat station for the RAF. Portland was visited for a week before the squadron reached Calshot.

One of the oddest events to occur to a flying-boat squadron resulted from the sale of a No. 201 Squadron's Southampton II (S1235) to Imperial Airways to replace their Short Calcutta flying-boat (G-AADN '*City of Rome*') lost off Spezia in a gale. Apart from being deprived of their boat, a 201 Squadron crew left on 26 November 1929 to deliver it to Athens where, as a commercial aircraft, it was accordingly registered as G-AASH.

Corrosion was always a problem in flying-boats and, at the end of January 1930, all Southamptons which had flown more than 300 hours were grounded. Certain stainless steel fittings had failed and replacements were slow to materialize. Exercises with the Fleet embraced several tasks. While one series would be concerned with hunting submarines, the next would require the flying-boats to locate a naval force and home a submarine flotilla on to it—foreshadowing German use of Focke-Wulf FW 200 Condors over Atlantic convoys in World War Two.

In September 1930, the squadron set off on another, more ambitious, cruise. Group Captain E. R. C. Nanson, CBE, DSC, AFC was attached to No. 201 to command the flight of three Mk. IIs and 2 wooden-hull Mk. Is which would tour the Baltic. With S1228 as flagship, S1229, S1234 and S1058 (a Mk. I) followed their leader across the North Sea to Denmark's Esbjerg and on to Copenhagen. Their route then took them to Stockholm, Helsinki, Tallinn, Riga, Memel, Putzig (in Poland), back to Stockholm and home via Göteborg, Oslo and Esbjerg. Exactly one month after departure, three boats arrived back at Felixstowe leaving one at the final stop awaiting replacement of a broken flying wire. For a month, the four flying-boats had maintained themselves away from base by their own efforts.

The squadron returned to Scotland in September 1931, visiting Oban which would become another flying-boat station during World War Two. Next summer No. 201 returned for a further visit and again in May 1933 when five of the Southamptons flew across to Londonderry in Ulster at the beginning of July to meet General Italo Balbo's fleet of 24 Savoia-Marchetti S.M.55X flying-boats on a cruise to the United States. The twin-hulled Italian monoplanes made an obvious impression on the RAF; from then on, every large formation of aircraft was described as a 'Balbo', after the Italian Air Minister.

With other squadrons setting-off on cruises to the Mediterranean, it was by now almost inevitable that No. 201 should undertake its 1934 cruise 'traditionally' to the Hebrides and Orkneys.

By the mid-1930s, Southamptons were an ageing breed and, in April 1936, a squadron crew collected their first Saro A.27 London I (K5257) from Saunders-Roe's works at Cowes, Isle of Wight, only a few miles across the Solent from Calshot.

The London was an all-metal biplane powered by two 820 h.p. Bristol Pegasus III radial engines mounted on the top wing. In place of the draughty cockpits to which No. 201's crews had become inured, Londons had a cabin for the pilots and accommodation within the hull for radio, galley and berths. Gun positions were fitted in the nose, dorsal and tail positions, the last-mentioned being mounted in the tail support to provide a wide field of fire astern. With a maximum speed of 145 m.p.h., the new boat had a range of over 1000 miles and could, if required, stay aloft for $13\frac{1}{2}$ hours with extra fuel.

The first production London (K5257) was joined in July 1936 by two more (K5259 and K5260); followed the next month by a fourth London I (K5261). At last, the Southamptons could be passed on. At the same time, Air Marshal Sir Arthur Longmore KCB, DSO visited to present No. 201's new squadron badge. It was a fitting occasion since Squadron Commander Longmore, as he then was, had been No. 1 Squadron's first Commanding Officer in the early days of naval aviation.

On 9 December 1936, two Southamptons (a Mk. I N9900 and a Mk. II S1232) were written-off-charge and the new year found the squadron preparing for a new cruise. For once, it was not to be to Scotland. Four London Is (K5259, K5260, K5262 and K5909) left Calshot on 15 January 1937 for Malta, using the French seaplane stations at Hourtin (near Bordeaux) and Berre (near Marseilles), returning at the end of the month via Algiers, Gibraltar, Lisbon and Hourtin.

During the first half of 1938, the squadron replaced its original Londons with improved London IIs, powered by 1000 h.p. Pegasus Xs. These it took to Invergordon, Ross and Cromarty, at the end of September 1938 as the Czechoslovakian crisis came to a head. After only 10 days, No. 201 returned to Calshot, the Munich agreement having given Europe a breathing

---

1 The Spit at Calshot showing the hangars, the control tower on top of the old Martello Tower and Swordfish of the Seaplane Training Squadron. A London of No. 201 Squadron taxies in the foreground (*Photo: Flight Int'l. 15130*)

2 A Southampton (S1645) is launched at Calshot. Rubber-gaitered ground personnel are unshipping the beaching wheels, one of which is held on the left. Note the retracted radio mast on the top wing (*Photo: Flight Int'l. 10185*)

3 No. 201 Squadron arranged for a formal squadron photograph in front of one of its Southamptons in 1934 (*Photo: Flight Int'l. 10194*)

1 A London (K5257) is launched from the Calshot slipway. This boat served with No. 201 between November 1936 and May 1938
(*Photo:* Flight Int'l. *15132*)

2 London 'Y/201' shows its three gun positions in nose, amidships and in the tail support
(*Photo:* Flight Int'l. *15129*)

3 'S/201' taxies into the mooring area to the north of Calshot Spit to join other Londons and the Scapas of the Seaplane Training Squadron
(*Photo:* RAF Kinloss *26969*)

4 Flight-Lieutenant Kendrick watches an engine change on a camouflaged London of No. 201 Squadron in November 1939. The squadron's code letter 'ZM' have been applied under the cockpit
(*Photo: No. 201 Squadron collection*)

space. However, during the following summer, the squadron surveyed alighting areas in the Orkneys and Shetlands accompanied by the transport *Manela*, the RAF's maid-of-all-work.

On 8 August 1939, six boats left Calshot for Scotland —to Stranraer, Wigtownshire—and three days later moved up to the Shetlands. Though the crews were not aware of it at the time, No. 201 Squadron had gone to war.

Northern Patrol

Even in summer, the squadron's new base looked bleak. The Shetland Islands could best be described as 'wind-swept' and flying-boats huddled in the relatively sheltered waters of Sullom Voe, one of several sea-lochs on the eastern shores of the cluster of islands. Nearby Garth Voe could also be used as a mooring area and occasionally flying-boats were based at Lerwick, the largest town in the Shetlands.

Huts began to be erected on shore but the head-quarters of No. 100 Wing—which controlled units in the Shetlands—was aboard the faithful *Manela* anchored in Sullom Voe. An airstrip was to be provided ashore at Scatsta; but until it was ready, communications by air operated into the civil airfield at Sumburgh—soon to be expanded into a busy Coastal Command base.

The Londons of Nos. 201 and 240 Squadrons were supplemented by the new Short Sunderlands detached from Southern bases. Their main task was to maintain a regular patrol across the Northern approaches to the North Sea, plugging the gap between Norway and the Shetlands. Through this stretch of ocean would pass German merchant ships trying to regain their home ports and, in the opposite direction, surface raiders and U-boats breaking out into the Atlantic.

No. 201's first wartime patrol began on 4 September with Squadron Leader Finlay (in K5260), Flying Officer Furlong (L7041) and Flying Officer Middleton (L7043) heading out on the long haul to the Norwegian coast. Further south, land-based Lockheed Hudsons from Leuchars were on a similar patrol, the gap between the two patrol lines ensuring that, in theory, anything missed during the night by one patrol would be picked up by the other—weather and aircraft serviceability permitting.

Several attacks were carried out on suspected U-boats without results. Day after day, the elderly flying-boats plied across the North Sea in all manner of weather. On 4 November, five Londons and two Sunderlands swept the patrol line without success for a London II (K9686) of No. 240 Squadron overdue with its crew of six. Two days later, No. 201's main base moved to Invergordon (later known as Alness) where mainten-ance was easier and communications more direct.

Sullom Voe was an isolated base. Its nearest fighter support was from the Royal Navy at Hatston in the Orkneys. Both groups of islands were favourite targets for German reconnaissance aircraft keeping watch on the Home Fleet in Scapa Flow. It was no surprise to see, on 10 November, an enemy bomber flying over the Voe, followed by a trail of exploding shells from cruisers moored nearby. Three days later, four more raiders dropped bombs in the water—and one near Sullom village—without causing any damage.

On the following day, Middleton and his crew were on patrol (in K5912) when a drop in oil pressure caused engine failure. The London II landed in the sea and its crew was picked up by the destroyer HMS *Imperial* which scuttled the flying-boat and took the crew to Methil, Fifeshire. A second London II (L7042) was lost on 22 November when, moored in Lerwick harbour and undergoing an engine change it was bombed and machine-gunned by six Heinkel He 111s; but without casualties to the crew. Damaged beyond repair, the hulk was towed into deep water and scuttled on 6 December.

The last patrol with a London was on 11 April 1940. Conversion to Short S.25 Sunderland Is began at Invergordon, four of the big monoplane flying-boats being taken over on 4 April to permit training to get under way. There was much to be learned. In place of the old biplanes, No. 201's crews had to get used to much larger boats with power-operated turrets for the gunners who had flown their patrols in open—and often freezing—cockpits.

Developed from the 'Empire'-class flying-boats built for Imperial Airways, the Sunderland was heavier and more powerful than its civil counterpart. Four 1000 h.p. Bristol Pegasus XXII radial engines gave the Sunder-land a speed of 210 m.p.h. and a patrol endurance of about $13\frac{1}{2}$ hours. Two turrets were fitted to early Sunderlands, one in the nose with one 0·303-in Browning machine-gun and the second forming the after end of the hull covering a wide field of fire with its quadruple Brownings. On top of the hull, two open gun positions each had a single 0·303-in Vickers K-gun covering beam angles and in later models these were replaced by a two-gun turret. A load of 2000 lb. of bombs or depth charges could be carried internally, racks being run out under the wing roots when required.

Most of No. 201's missions were long patrols over the North Atlantic; its main enemy the U-boat. In addition, however, Sunderlands were on call for reconnaissance flights to areas outside the range of land-based aircraft. Northern Norway was one of these regions and, on 20 September 1940, Flying Officer Lindsay and his crew (in 'U/201'; L5802) reconnoitred Narvik. Taking-off in the late evening, the Sunderland made landfall off the Andoy Islands and headed up the coast to Narvikfjord. Two new camps were located at Bjervik and an oil storage tank at Harstad was bombed. Unfortunately, only one of the three bombs selected released. After checking shipping in the area, the Sunderland I landed back at Sullom Voe after a flight of $14\frac{1}{2}$ hours.

A similar mission on 21 January 1941 almost ended in disaster. A Sunderland I ('O/201'; T9049), with Squadron Leader Cecil-Wright and crew aboard, set out in the early hours of the morning and arrived off Namsos five hours later. Bodo harbour and airfield were investigated before heading north to Bjervik where a bomb was dropped on a German camp. After running up to Narvik through Beisfjord, three more bombs were dropped—on a camp and also a 6000-ton ship—at Fagernes Point. Heavy and light anti-aircraft guns replied, hitting the tail and severing hydraulic lines to both turrets. Course was set for the Shetlands but, because of a snow-storm in the area, an emergency landing had to be made at Herma Ness some 15 hours after take-off. With the port wing and float badly damaged in landing, the crew were then forced to climb out on to the starboard wing to balance the boat while Squadron Leader Cecil-Wright taxied into Woodwick. Here, the damaged Sunderland drifted on to rocks before it could be secured. The crew scrambled ashore and a few days later 'O/201' was towed to Culli Voe for salvage. In August 1941, it rejoined No. 201 again.

One of the few squadron boats to encounter an enemy aircraft during this phase of the war was flown by Flight Lieutenant Lindsey ('Y/201'; L5805). He was on patrol between Iceland and the Faeroes when a Focke-Wulf FW 200 Condor four-motor reconnaissance-bomber was sighted. Jettisoning all bombs and depth-charges, the Sunderland opened fire at 400 yards range but after the initial attack was outrun by the Condor which made off. Distress calls from a Condor were later picked up by RAF Coastal Command monitoring stations as a luckless FW 200 'ditched' near Bordeaux. This may have come from the same aircraft engaged by 'Y/201'.

Flight Lieutenant Vaughan's crew (in 'Z/201'; L5798) took-off late on 23 May 1941 with orders to locate and shadow enemy warships passing through the Denmark Strait between Iceland and Greenland. While searching the area, gun flashes were observed in the distance. Moving closer, four ships could be seen firing, one of which blew up in a column of black smoke; it was later found to have been the World War One battle-cruiser HMS *Hood*. Two ships heading away from the scene of the action were identified as the new German battleship *Bismarck* and the *Admiral Scheer*; though the

1 Supermarine Southampton S1249 served with No. 201 Squadron from 25 August 1930 until handed over to the Seaplane Training Squadron on 4 June 1932. The squadron badge was carried on the nose forward of the large identification number.

2 Sopwith Camel F3227 carried No. 201's single vertical identification bar aft of the fuselage roundel and an individual letter, conforming with the former Royal Flying Corps practice.

3 Saro London II K5257 was collected from the factory of Saunders-Roe Ltd on 22 April 1936 and served with No. 201 until May 1938. Transferred in turn to Nos. 204 and 240 Squadrons and No. 4 (Coastal) Operational Training Unit, this boat was finally struck off charge on 13 November 1942.

© Hylton Lacy Publishers Limited

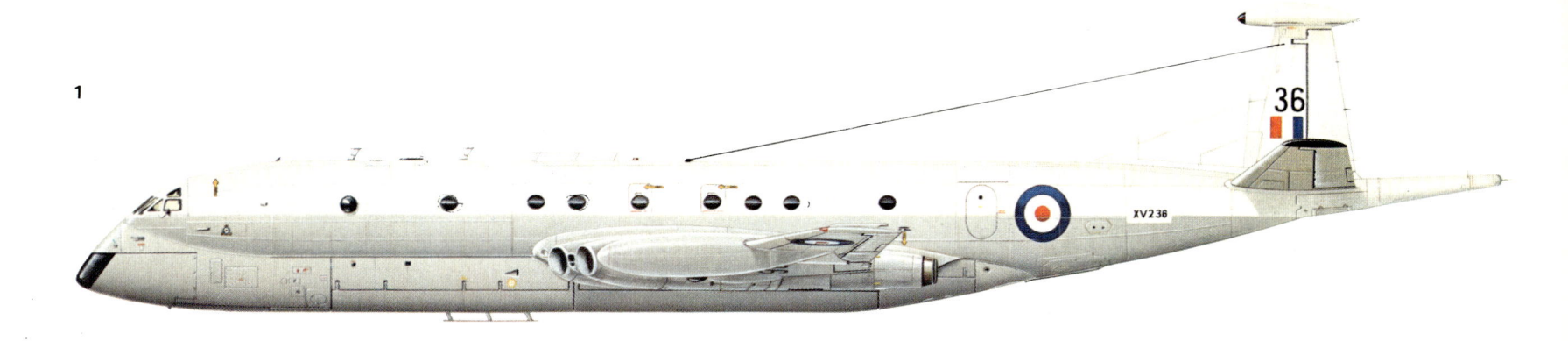

1 Hawker Siddeley Nimrod MR.1 XV236 in the standard Strike Command maritime colour scheme as part of No. 18 (Maritime) Group. Individual identification is effected by reproducing the final two digits of the serial number above the fin flash.

2 Short Sunderland MR.5 RN284 was built by Blackburn at Dumbarton in 1945 and after a period with No. 201 from June 1945 to May 1947 served with No. 230 Squadron, No. 235 Operational Conversion Unit and the Flying Boat Training Squadron before returning to No. 201 in June 1955 for service until the squadron's disbandment, when the aircraft was sold to Messrs. Short Bros. & Harland at Belfast.

3 Shackleton MR.3 XF707 served with both Nos. 201 and 206 Squadrons at Kinloss. With No. 201, it was coded 'P' and with No. 206 as 'C.'; it is depicted here in its No. 206 markings which incorporated the squadron badge.

© Hylton Lacy Publishers Limited

1·2

3·4

5·6

36

1 Sunderland 'W/201' over choppy water in October 1941. ASV radar arrays are mounted on top of the fuselage and under each wing
(*Photo: No. 201 Squadron collection*)
2 Icebound on Lough Erne Sunderland 'Q/201' (ML742) during the unprecedented freeze-up in January/February 1945
(*Photo: No. 201 Squadron collection*)
3 Engine servicing at a Sunderland mooring at Castle Archdale. Leading-edge platforms could be let down to form working platforms but all tools had to be firmly anchored to the aircraft in case of accidental droppage. The turret is retracted to make room for a mooring gallery and bollard
(*Photo: IWM. CH11073*)
4 The Sunderland V that carried out No. 201's last patrol in World War Two takes off from the waters of Lough Erne (*Photo: IWM. CH15361*)
5 After the end of the war, No. 201's boats received their final colour scheme. Squadron badges appeared below cockpits and single-letter codes were adopted for squadron identification
(*Photo: Flight Int'l. 32858*)
6 Symbolic of thousands of similar convoy patrols, 'Z/201' flies over its charges in the North Atlantic
(*Photo: IWM. CH15302*)
7 A U-boat under attack in the Atlantic. Machine-gun fire sprays the submarine to suppress return *flak* during the run-up but the U-boat has already started to crash-dive
(*Photo: IWM. C2611*)
8 'Z/201' (DD829) carries out a night attack on a U-boat in the English Channel on 7 June 1944 as the second day of the invasion of Europe comes to an end (*Photo: IWM. C4599*)

latter was in fact the heavy cruiser *Prinz Eugen*. The battleship *Bismarck* was trailing oil and smoke and a pair of Royal Navy County-class cruisers were shadowing her on each flank. After three hours over the ships, the Sunderland had to leave for home and subsequent searches failed to locate the enemy.

Fortunately, a Consolidated Catalina later found the giant battleship and it was brought to action by the Home Fleet and Force 'Z' from Gibraltar and sunk before it could escape to Brest.

Not all the danger came from the enemy. Convoys occasionally fired on escorting Sunderlands and on 31 July 1942, 'P/201' (W4025) was shot down by a ship in convoy WS21 which it was escorting. Next day, another Sunderland was lost when 'R/201' (W4000) was seen to land five miles from the same convoy. Eye-witness reports stated that after floating for two minutes, it was seen to blow up; with the loss of the entire crew. During March 1943, several attacks were made on U-boats without sinkings being confirmed but on 31 May, No. 201 Squadron had its first success. Flight Lieutenant Hall and crew aboard a Blackburn-built Sunderland III ('R/201'; DD835), sighted a surfaced U-boat and dropped four depth charges on its wake as it submerged. The loss of *U-440* was later confirmed.

A new weapon in the sea war was encountered in November 1943 when a Sunderland III ('O/201'; DD860) captained by Squadron Leader W. D. B. Ruth, sighted a four-motor Heinkel He 177 heavy bomber approaching the convoy it was escorting. Cutting across its path, the Sunderland's gunners opened fire at 400 yards. The Heinkel then jettisoned its novel Henschel Hs 293 radio-controlled, rocket-powered glider-bomb and made off. Ten minutes later a Focke-Wulf FW 200 Condor was chased off the scene but a second succeeded in launching its bomb before the Sunderland could reach it. Before reaching the limit of its fuel, yet another He 177 was engaged by 'O/201' which then headed back for Castle Archdale in Fermanagh, Ulster.

Two days after Christmas 1943, Flight Lieutenant Baveystock and crew (in 'T/201'; EJ137) were searching for a blockade runner in the Bay of Biscay when the German raider *Alsterufer* was sighted and shadowed. A low run over the ship with turrets raking the deck resulted in a well-aimed or lucky cannon shell arriving in the bomb room. In reprisal, two depth charges and two 500-lb. bombs were dropped on the raider while other Coastal Command aircraft were 'homed' on the Sunderland's position. Fuel was running short when 'T/201' headed back to England and a landing was made off the Scillies. The flying-boat was towed into St Mary's with only 60 gallons left in the tanks. Flight Lieutenant Stack (in 'U/201'; EK579) relieved 'T/201' and guided a Czechoslovak-crewed Consolidated Liberator to the spot. The four-motor Liberator sank *Alsterufer*. Stack returned to Castle Archdale: having been 18 hours 35 minutes in the air.

During March 1944, preparations for a move to Pembroke Dock in Wales began and, early in April, the squadron was ready to operate off the Bay of Biscay and the approaches to the Channel. D-Day was approaching and one of RAF Coastal Command's main tasks would

37

be to prevent the U-boats and small surface vessels based on the Biscay ports from reaching the vast armada of invasion craft gathered off the Normandy beaches. By this time, No. 201's Sunderlands were better armed than ever before. As many U-boats remained on the surface when sighted by aircraft, relying on their anti-aircraft armament instead of seeking the now-illusory security of submergence, *flak*-suppressing armament had been fitted. Five fixed 0·303-in Browning machine-guns supplemented the Browning in the Frazer-Nash FN 5 nose turret. There were two more Brownings in the mid-upper turret and four more in the tail turret. Two 0·303-in Vickers K-guns were mounted in the galley, one on each side of the hull, to cover beam positions. Eight 250-lb. Torpex-filled Mark XI depth-charges had a damage radius far in excess of earlier anti-submarine bombs and a special low-level bomb-sight was fitted.

On the evening of D-Day, Baveystock was airborne (in 'S/201'; ML772) over the western approaches to the Channel. Five minutes before midnight, the ASV Mark III radar picked up an echo at nine miles range and the flying-boat began a target run-in at 250 feet. Half-a-mile from the contact, the trace disappeared from the screen. Flares were dropped which revealed a swirl and bubbles. Nearly three hours later, another trace appeared on the radar screen which was tracked. Light *flak* began to come up at the Sunderland and flares illuminated a fully-surfaced U-boat. All forward guns returned the fire and six depth charges (followed by a marker) dropped towards the enemy submarine. The explosions shook the flying-boat as the charges straddled the U-boat. After an hour of patrolling in the vicinity, no further trace appeared on the radar. *U-955* as the German submarine was later identified, had gone to the bottom.

Four nights later, the same 'S/201' was not so lucky. A Vickers-Armstrongs Wellington of No. 172 Squadron saw tracers in the distance followed by an aircraft on fire. No further signs of 'S/201' were found and Squadron Leader W. D. B. Ruth, DFC and his crew were posted missing. 'P/201' (ML881) under the command of Flight Lieutenant I. F. B. Walters, DFC, located a U-boat when the second pilot, Flight Lieutenant S. C. Buzzard, sighted a wake from 8½ miles away. The look-outs aboard *U-1222* were not as sharp as the 'Eyeballs Mk. I' on board 'P/201' as it began to dive when the Sunderland was only half-a-mile away. Five depth charges lifted the submarine and debris marked its end.

Despite the large numbers of U-boats in commission, only one Allied supply ship was sunk by submarines during the invasion period—an indication of the degree of success attained by Coastal's patrol squadrons. Baveystock's crew (in 'W/201'; EJ150) sank *U-107* off St Nazaire on 18 August 1944 by placing six depth charges round a periscope wake. Flight Lieutenant D. R. Hatton ('Y/201'; ML882) was directed to a position west of Ireland on 6 December only to arrive to find a frigate rescuing survivors from a U-boat. Five miles away, smoke from a '*Schnorkel*' air-intake exhaust outlet betrayed the position of *U-297*. On the first run-in, none of the depth charges released but on the second all six landed just ahead of the wake. An oil slick appeared and spread while escort vessels found no echo from their Asdic submarine-detection devices. Though the British Admiralty reported that the U-boat had reached harbour, their intelligence was on this occasion wrong—*U-297* had been sunk.

For the final months of the war, No. 201's Sunderlands patrolled with only one definite success. Flight Lieutenant K. H. Foster and crew (in 'H/201'; ML783) attacked *U-325* with gunfire but yet again the depth charges failed to release. A marker was dropped and a sighting report made which resulted in the destroyers HMS *Hesperus* and *Havelock* sinking the U-boat. The end of the Battle of the Atlantic was near and on 9 May 1945, the crew of 'V/201' (ML764) sighted *U-1105* flying a black surrender flag. Two days later, the same aircraft dropped depth charges on a '*Schnorkel*'. All U-boats had been ordered to surface and RAF Coastal Command were taking no chances. Finally, on 3 June, Wing Commander J. Barrett lifted a Sunderland III ('Z/201'; ML778) off the waters of Lough Erne (Castle Archdale) on No. 201 Squadron's last operational patrol. It was also the last wartime patrol of Royal Air Force Coastal Command.

Maids-of-all-Work

As World War Two came to an end in Europe, No. 201 Squadron completed its re-equipment with Sunderland Mark Vs. The old Pegasus engine (which had powered all previous Sunderlands marks) had managed to cope with increased weights, long periods of continuous running and high power settings but there was a limit to its development. Two Sunderlands were re-engined with 1200 h.p. Pratt & Whitney R-1830 'Twin Wasps' for trials, one by Short Bros. at Rochester and the other by No. 10 Squadron, Royal Australian Air Force, at Mount Batten. The success of this installation resulted in all future production aircraft being fitted with R-1830s and several Mark IIIs were converted during overhauls.

The planned replacement for the Sunderland III was the Sunderland IV (later renamed Seaford) powered by Bristol Hercules XIXs of 1500 h.p. each. The mid-upper turret was to mount a pair of 20-mm Hispano cannon, and the nose and tail turrets each had two 0·50-in Browning heavy machine-guns. Two more Browning heavy machine-guns were in beam positions and a pair

---

1 The Sunderland's 'secret weapon'—a galley that provided genuine, freshly-cooked, hot meals during long patrols (*Photo: IWM. CH11086*)

2 The crew of 'Z/201' (ML778) poses for photographs at Castle Archdale before embarking for Coastal Command's last wartime patrol on 4 June 1945. Dark uniforms are worn by Royal Australian Air Force personnel (*Photo: IWM. CH15304*)

3 A trio of Sunderland GR.5s of No. 201 Squadron formate over the coast of Pembrokeshire (*Photo: Flight Int'l. 32862*)

4 Pembroke Dock mooring area in the early 1950s. The Sunderlands of Nos. 201 and 230 Squadrons are attended by the local marine craft section—refuellers, seaplane tenders and a fire boat (*Photo: Flight Int'l. 32856*)

5 An embarrassing moment for 'E/201' on a courtesy call to Guernsey. On 15 September 1954, this Sunderland (RN271) hit a rock and was towed into harbour in a sinking condition. Salvaged, 'E/201' was eventually scrapped three years later (*Photo: via Peter M. Corbell*)

6 Escorted by two of No. 201's Shackletons, the last Sunderland comes home to Pembroke Dock on 24 March 1961. Formerly 'Z/201', ML824 had been transferred to the French Navy in October 1951 and served with Aéronavale *Flottilles* 7F, 12S, 27F and 50S before being presented to the Sunderland Trust for preservation. This boat is now exhibited in the Royal Air Force Museum, Hendon (*Photo: No. 201 Squadron collection*)

---

of smaller 0·303-in machine-guns were fixed in the nose.

In April and May 1946, No. 201 carried out operational trials on the Seaford in co-operation with the Marine Aircraft Experimental Establishment at Felixstowe. Hopes of being re-equipped with the improved boats faded as the Seaford programme was abandoned for reasons of economy and the existing aircraft converted to Solents for commercial use.

Ulster's Castle Archdale was abandoned as a permanent base soon after the end of the war though it continued to be used for exercises. No. 201 flew its boats back to Calshot in formation in March 1946 after eight months at Pembroke Dock. Training and ferry flights went on for more than two years before a new task was given to the squadron. On 2 July 1948, Nos. 201 and 230 Squadrons were at Castle Archdale on exercises when an urgent recall signal brought them back to Calshot. Next day, the Elbe near Hamburg was dotted with white Sunderlands.

The reason was the Russian blockade on Berlin, imposed to force the other three Occupying Powers to

39

evacuate their zones of responsibility in the former German capital. Having agreed to Russian control of the access roads and railways at the end of the war, the British, American and French zones of Berlin now found themselves cut off from the rest of their occupation zones. Within a few days, the Russians were confident that the economic life of West Berlin would grind to a halt. They assumed also that the Western Powers would not permit the population to starve and viewed the first aircraft to fly into Tempelhof Airport with supplies as merely a gesture.

The massive transport fleets which had carried an airborne army into Europe, supplied another deep in the Burmese jungle and spanned the world with air routes to provision the ground and naval forces, had been dispersed. But from far and wide, the US Air Force and the RAF gathered Douglas C-54/DC-4 Skymasters, Avro Yorks, Douglas C-47/DC-3 Dakotas and any other transports they could press into service. With their roomy hulls, Sunderlands were an obvious reinforcement for conventional transports.

On 5 July 'X/230' took off from the former Blohm und Voss AG factory at Finkenwerder on the first sortie to the Havel See in Berlin. It was followed by many more from Nos. 201 and 230 Squadrons and from the training boats of No. 235 Operational Conversion Unit. Operating from the congested and dangerous waters of the Elbe, the Sunderlands embarked 10,000-lb. loads from amphibious trucks and took off amid a maze of small craft, wrecks and sandbanks. At the other end, the big flying-boats had to fly low over the city to land on the lake with an ever-present danger of hitting logs dropped by the Russians into the waterways entering and passing through the lake. These enabled Russian-controlled craft to pass right through the landing area and presented an ideal opportunity for sabotaging the operation and causing accidents.

Since flying-boats were designed to operate at sea, corrosion was not so great a problem as in landplanes. Consignments of salt were therefore allotted to Sunderlands as well as a variety of other goods. During the 12-minute turn-round at the Havel See, manufactured goods from the factories of Berlin were loaded for transfer to Western Germany. During November, the flying-boats evacuated large numbers of children and invalids as winter threatened the beleaguered city. Many days were foggy but the airlift continued through December; but floating ice on the Elbe and Havel See made operations by Sunderlands too hazardous. On 15 December 1948, the flying-boats ended operations but the Berlin Airlift went on. By the middle of February 1949, the one millionth ton of freight reached Berlin; and, on 12 May, the Russians gave up and ended their blockade. The airlift continued until October, the

1 Castle Archdale's maintenance area at the end of World War Two. Visible are 12 Sunderlands and three Catalinas of Nos. 201, 202 and 423 Squadrons *(Photo: No. 201 Squadron collection)*

2 Two of 201's Sunderlands take-off in formation from Milford Haven *(Photo: Flight Int'l. 32854)*

3 Shackleton 'N/201' over Pentire Point, Newquay, Cornwall, carries the squadron emblem on the fin *(Photo: MoD. PRB26060)*

4 A sonobuoy and torpedo leave a Shackleton MR.3 of No. 201 Squadron during exercises *(Photo: No. 201 Squadron collection)*

two millionth ton being flown in soon after the first anniversary of the operation. If the Russians had regarded the reopening of land routes as an interim measure to encourage the dismantling of the complex air transport structure so that the blockade could be reimposed, they were disappointed. When the RAF flew its 65,857th and last flight of '*Operation Plainfare*' in October 1949, Berlin was out of danger.

In July 1951, four Sunderlands set out on a cruise to the West Indies; while another departed for less accessible shores. A British expedition to North Greenland required transportation to a point only 700 miles from the North Pole. On 23 July Wing Commander Barrett flew the party to Iceland and, two days later, took the explorers on to a lake near the site of their investigations. On the return flight, the Sunderland brought back a sick infant from the base of a British scientist in Greenland, landing at Tayport, Fifeshire, to deliver the patient to Dundee Royal Infirmary.

The party was due to be picked up about four weeks later and Barrett left with supplies on 23 August but was weatherbound in Iceland for four days. Ice began to form on the lakes in North Greenland but the Sunderland eventually got through to find the five scientists with almost no food left. A year later, a British North Greenland Expedition on a much larger scale was planned to be airlifted into Britannia Lake, which had been located as a suitable base by No. 201's Sunderland. However, No. 230 Squadron was given the task of supporting the expedition in 1952. But, after spending the winter in Greenland, the isolated community eagerly awaited the melting of the ice on the lake and on 12 August, 'A/201' touched down. Five Sunderlands made flights into Britannia Lake to build up supplies for the winter and the operation was almost complete when 'C/201' (RN184) ran into trouble.

The roar of starting engines caused an icefall which sent ice floes drifting towards the Sunderland's taxi-ing area. In attempting to avoid them, 'C/201' ran aground and sank by the bow. Pumping and the use of a rubber dinghy as a bumper prevented the Sunderland sinking and after 'B/201' had flown in emergency kits, leaks were plugged in freezing conditions and the hull made watertight with concrete. For a while it looked as though the Sunderland would have to be laid-up until the following summer. But, at the last moment, 'C/201' was started-up and taxied out to the remaining stretch of clear water. As the boat churned across the lake, a gash opened up but 'C/201' became airborne and headed for Iceland with water streaming out of the hull. Throughout the flight, lanolin-impregnated cotton waste was added to the makeshift repairs. When, eventually, on 20 August, the Sunderland arrived at the maintenance base at Wig Bay, near Stranraer, its scars regarded with awe, where it was hauled up the slipway.

With the Sunderlands rapidly approaching the end of their fine careers, a replacement was urgently required and several designs were proposed. In the event, no replacement flying-boat was ordered and Nos. 201 and 230 Squadrons at Pembroke Dock were the last of the British-based flying-boat units. With a normal strength of 10 boats, the two squadrons retained one boat on permanent air-sea rescue stand-by while the others engaged in exercises with the Allied naval forces. On 16 December 1955, the Air Officer Commanding No. 19 Group, Coastal Command, presented the squadron with its Standard at a ceremony at Pembroke Dock.

The end came on 28 February 1957, when the squadron was finally disbanded and the waters of Milford Haven no longer echoed to the roar of Sunderland 'Twin Wasp' engines as the big boats lifted-off and headed out to sea.

All Ashore

When No. 201 Squadron reformed on 1 October 1958, it was at the vast RAF Coastal Command airfield at St Mawgan in Cornwall. Long concrete runways had replaced the wind-swept waters from which 201 had previously operated but what was lost in nautical know-how was more than made up for by operating from dry land. Tools (and, occasionally, airmen) no longer ended up in the sea but could be retrieved from the tarmac; even though, for a few of the less-surefooted, the concrete proved less yielding.

The squadron's new equipment was the **Avro** (latterly Hawker Siddeley) Shackleton MR.3, a patrol aircraft developed at the end of World War Two and powered by four 2450 h.p. Rolls-Royce Griffon piston engines driving contra-rotating propellers. The Shackleton was to have a crew of 10 and a range of over 4000 miles. Mounted in the nose was a pair of 20-mm cannon. Depth-charges, homing torpedoes, sonobuoys or auxiliary fuel tanks were capable of being housed in a vast bomb-bay. The Mark 3 differed from earlier marks in several ways, notably the provision of a tricycle undercarriage and the fitment of wingtip tanks. Later, two additional Viper jet engines were to be fitted in the inboard nacelles to improve the performance.

No. 201 did not have to go through the lengthy procedure of acquiring crews and aircraft. No. 220 Squadron was merely renumbered to bring the more senior squadron back into the line. Shipping reconnaissance, air-sea rescue missions and Nato-sponsored exercises filled the squadron's time. Then, in March 1965, No. 201 moved to Kinloss on the shores of the Moray Firth to guard the northern flank of the British Isles.

In October 1970, the squadron received new aircraft. During the previous month, the Shackletons were returned to maintenance units and their crews underwent a conversion course on the Hawker Siddeley Nimrod at No. 236 Operational Conversion Unit, St Mawgan. In place of the vibrating piston-engined Shackletons came the smooth four-jet development of the famous de Havilland Comet airliner. At first glance there was little resemblance between the sleek Comet and the bulky Nimrod. A lower deck had been built on to house radar and bomb-bay while the fin was topped by a radar housing. Though not handsome, but nevertheless sleek, the Nimrod was loaded with sophisticated equipment and its high cruising speed could take it rapidly to the assigned patrol area where power combinations could be effected to permit 'loitering' until a contact was picked up.

The combination of long range, high speed when required, and good handling for low-level searches has made the Nimrod probably the most efficient anti-submarine and maritime reconnaissance aircraft in the world.

From Kinloss, where a round-the-clock watch is kept on Soviet ships operating off the coasts of Nato member countries, a Nimrod set out on 3 January 1971 for a rendezvous with history over Alum Bay at the western tip of the Isle of Wight. There, over the waters that had been the training ground for the squadron's early flying boats, were dropped the ashes of Air Chief Marshal Sir Arthur Longmore who, as Squadron Commander Longmore of the Royal Naval Air Service, had gathered together the first men of 'Number One Naval'.

1 Three Nimrods on the apron at Luqa, Malta, during Exercise *Lime Jug* in 1971 (*Photo: MoD. TN6293/177*)

2 Nimrod 45 (XV245) takes-off from the Canadian Armed Forces base at Greenwood, Novia Scotia, in June 1971 after a courtesy visit to Canada (*Photo: RAF Kinloss 3703G*)

3 Shackleton MR.3 of No. 201 over the Cornish countryside while the squadron was based at St Mawgan (*Photo: MoD. PRB24606*)

4 A flight of four Shackletons in formation along the rocky Trucial coast during a detachment to The Gulf (*Photo: No. 201 Squadron collection*)

SQUADRON BASES

Fienvillers, France	1 April 1918
Noeux, France	12 April 1918
Ste Marie Cappel, France	20 July 1918
Poulainville, France	6 August 1918
Noeux, France	14 August 1918
Baizieux, France	19 September 1918
Beugnatre, France	14 October 1918
La Targette, France	27 October 1918
Bethencourt, France	22 November 1918
Lake Down, Wiltshire	15 February 1919 to 31 December 1919
Calshot, Hampshire	1 January 1929
Invergordon, Ross and Cromarty	29 September 1938
Calshot, Hampshire	7 October 1938
Sullom Voe, Zetland, Shetland Islands	9 August 1939
Invergordon, Ross and Cromarty	6 November 1939
Sullom Voe, Zetland, Shetland Islands	26 May 1940
Castle Archdale, Fermanagh, Ulster	9 October 1941
Pembroke Dock, Pembrokeshire	8 April 1944
Castle Archdale, Fermanagh, Ulster	3 November 1944
Pembroke Dock, Pembrokeshire	2 August 1945
Calshot, Hampshire	30 March 1946
Pembroke Dock, Pembrokeshire	18 January 1949 to 28 February 1957
St Mawgan, Cornwall	1 October 1958
Kinloss, Morayshire	14 March 1965

SQUADRON EQUIPMENT
Period of Use and Typical Serial and Code Letters

Sopwith Camel	April 1918 to February 1919	F5941 (E)
Sopwith Snipe	October 1918	E8102
Supermarine Southampton II	January 1929 to December 1936	S1235
Saro London I	April 1936 to June 1938	K5259
Saro London II	January 1938 to April 1940	K5257 (ZM-Y)
Short Sunderland I	April 1940 to January 1942	P9606 (ZM-R)
Short Sunderland II	May 1941 to March 1944	T9083 (ZM-Q)
Short Sunderland III	January 1942 to June 1945	ML743 (ZM-A)
Short Sunderland V	February 1945 to February 1957	SZ571 (201-D)
Hawker Siddeley (Avro) Shackleton MR.3	October 1958 to September 1970	XF708 (201-O)
Hawker Siddeley (D.H.) Nimrod MR.1	October 1970 to date	XV246

COMMANDING OFFICERS

Major C. D. Booker	1 April 1918
Major C. M. Leman	18 August 1918
S/Ldr D. G. Donald	1 January 1929
S/Ldr E. F. Turner	20 January 1930
S/Ldr C. G. Wigglesworth AFC	8 May 1933
S/Ldr J. D. Breakey DFC	1 April 1935
W/Cdr J. H. O. Jones	20 September 1937
W/Cdr C. H. Cahill AFC	1 February 1939
W/Cdr C. S. Riccard	18 August 1940
W/Cdr W. G. Abrams	15 April 1941
W/Cdr J. L. Crosbie	23 December 1941
W/Cdr J. B. Burnett	19 August 1942
W/Cdr R. E. G. van der Kiste	4 May 1943
W/Cdr K. R. Coates	15 July 1944
W/Cdr J. Barrett	22 February 1945
W/Cdr J. W. Louw	26 August 1945
W/Cdr W. H. Tremear	21 March 1946
W/Cdr J. L. Crosbie	5 May 1947
S/Ldr D. H. F. Horner	1 September 1947
S/Ldr R. C. L. Parkhouse	27 January 1949
S/Ldr H. A. S. Disney	7 February 1950
S/Ldr P. A. S. Rumbold	4 December 1950
S/Ldr R. A. N. McCready	17 November 1952
S/Ldr D. W. Bedford	3 January 1955
W/Cdr J. G. Roberts	1 October 1958
W/Cdr A. C. Davies	18 August 1959
W/Cdr R. B. Roache	6 September 1961
W/Cdr P. G. South	1 July 1963
W/Cdr W. S. Northcott	9 April 1965
W/Cdr N. Jones	23 November 1966
W/Cdr G. A. Chesworth	30 August 1968
W/Cdr J. B. Duxbury	1 March 1971
W/Cdr J. M. Alcock	8 October 1971

202 Squadron

1 This D.H.4 (N5997) was built by Westland for the RNAS and was on the strength of No. 2 Squadron, RNAS when it became No. 202 Squadron. The elaborate markings soon succumbed to the RAF's desire for anonymity in its aircraft (*Photo: MoD .H1536*)

2 Fairey IIIDs formed the squadron's equipment when No. 481 Flight was renumbered No. 202 Squadron (*Photo: MoD. H1340*)

3 The D.H.4 was No. 202's standard type in World War One. The gap between the pilot and gunner seen in this view was a weakness in this design as it impeded communication and made co-ordination under attack difficult (*Photo: IWM. Q68130*)

Over the Belgian Coast

When No. 202 Squadron, Royal Air Force, was formed on 1 April 1918, it inherited the traditions of 3½ years of service as No. 2 Squadron, Royal Naval Air Service. Born on 17 October 1914, No. 2 Squadron was one of two squadrons approved for formation with a view to co-operating with the Royal Flying Corps in France. Squadron-Commander E. L. Gerrard began to assemble his unit at Eastchurch, Kent, and started training with Bristol T.B.8 biplanes transferred from the Military Wing.

On 24 December 1914, two Bristol biplanes (Nos. 1223 and 1224) flew down to Dover for air defence duties. On Christmas Day, No. 1224 was ordered up after an enemy aircraft was reported over the Medway. Over Herne Bay the pilot sighted the enemy a long way off but failed to intercept. However, it was No. 2's first sight of its enemy.

Replaced by Avros of No. 1 Squadron, the detachment returned to Eastchurch on 2 January 1915. By this time the squadron had a collection of varied aircraft including a Vickers F.B.5 (No. 32), an Avro 500 (No. 53), a Short S.28 (No. 66), a Sopwith biplane (No. 104), three Bristol T.B.8s (Nos. 1223 to 1225) and even the Royal Navy's first numbered aircraft, (No. 1), a Short S.34.

During February 1915, even more types were added to the motley collection; an Avro 503 (No. 16) seaplane converted for land use, a B.E.8 'Bloater' (No. 643) transferred from the RFC and a French-built Blériot parasol monoplane (No. 908). These, augmented by the three T.B.8s, were despatched to Dover on 10 February to take part in bombing raids on Zeebrugge and Ostend.

To guard south-eastern England against enemy airship raids, a detachment of No. 2 Squadron ('A' Flight) was sent to Westgate at the end of April. One pilot from Westgate, Flight Sub-Lieutenant R. H. Mulock, intercepted Zeppelin *LZ38* on 17 May 1915. Caught cruising around Kent at 2000 feet, *LZ38* hastily dropped bombs on Ramsgate and rapidly climbed out of range, far too fast for Mulock's Avro to catch up. After a chase almost to the Belgian coast, the Zeppelin passed out of sight.

During June 1915, RNAS Squadrons were re-designated 'Wings' and during August, No. 2 Wing spent a short time at Dunkerque. A week after arrival, Flight-Commander J. R. Smyth-Piggott in a B.E.2c succeeded in bombing Zeppelin *L12* as it was being towed into Ostend by German torpedo-boats after being damaged by anti-aircraft guns at Dover. Other pilots from Dunkerque followed and the airship was damaged beyond repair. A few weeks later, No. 2 Wing was on its way to the Dardenelles.

On 1 March 1916, No. 1 Wing, RNAS at Dunkerque, was divided into 'A' and 'B' Squadrons; and, on 5 November 1916, 'B' Squadron became No. 2 Squadron, RNAS. Equipment during this period was mixed. For spotting duties using radio to communicate with monitors and the old pre-Dreadnought *Revenge* bombarding the Belgian coast Farman F.40s were used. They were backed-up by a pair of B.E.2cs and given cover by Nieuport 10 and Sopwith Pup single-seat fighters. Also on hand in small numbers were Sopwith 1½ Strutter two-seaters which could be used for escort or reconnaissance duties.

Enemy aircraft based on Belgian airfields made many attempts to interfere with reconnaissance and spotting aircraft. Therefore, it was with some relief that the squadron received the first of its de Havilland (Airco) D.H.4s on 6 March 1917 when the prototype (No. 3696) arrived at St Pol followed slowly by production aircraft. By October 1917, No. 2 had 15 D.H.4s (variously powered by three different types of engine) and six Sopwith 1½ Strutters (with two types of Clerget rotaries in use). By the end of November, full equipment with D.H.4s was complete.

The translation of No. 2 Squadron, RNAS, to No. 202 Squadron, RAF, on 1 April 1918 meant little change in role. Its sphere of operations was still the Belgian coast and reconnaissance and spotting remained the primary tasks. During May 1918, four D.H.9s were received but returned to the Aircraft Depot within a short time; even though the D.H.9 had been designed to replace the D.H.4, most crews preferred their original aircraft. Despite opposition from enemy fighters, the D.H.4s still got through to their objectives even when operating singly or in pairs.

At the end of September 1918, bombardment of the Belgian coast was stepped-up and the Belgian and British armies in Flanders moved forward during October from the lines they had held for four long years. The Germans began to evacuate Belgium rapidly and the squadron was busily engaged in trying to locate the position of his rearguards and verify the towns which had been abandoned. Some methods were unorthodox as when a Camel of No. 210 Squadron 'captured' Ostend by landing in the market place. Varssenaere airfield was occupied by the advancing troops and No. 202 moved in to shorten the range between it and the German armies retiring over the Meuse a few days after the Armistice.

In February 1919, No. 202 was reduced to a cadre as a result of the demobilization of many RAF personnel. The remaining members of the squadron moved to Eastburn airfield, later to become famous as Driffield, in Yorkshire to await disbandment as part of No. 16 (Training) Group. During March, four of its faithful D.H.4s were passed to No. 233 Squadron at Dover and 10 more to No. 98 Squadron at Alquines, France. The end came on 22 January 1920 and No. 202 dispersed.

In the Mediterranean

No. 202 did not disappear for long from the RAF List. On 9 April 1920, it was reformed at Alexandria as a Fleet co-operation unit. Until the RAF could be disposed of tidily, the Royal Navy had, for want of an alternative, to work with the new service. On their side, the RAF had agreed to provide a number of squadrons to work with the Fleet and it was logical to allot former ex-RNAS squadron numbers to these units. No. 202's new role was short-lived and lasted just over a year. Small budgets, lack of personnel and little interest in the Royal Navy in shore-based aircraft resulted in No. 202 being disbanded on 16 May 1921. For nearly eight years, maritime aviation in the Mediterranean would be based on the seaplane station at Malta and its resident unit, No. 481 Flight.

A resurgence of the flying-boat squadrons occurred in January 1929. Existing flights were raised to squadron status and on the first day of 1929, No. 481 (Coastal Reconnaissance) Flight became No. 202 Squadron. More precisely, it became No. 202 (Flying Boat) Squadron, a blatant misdescription since the available aircraft consisted of six Fairey IIID seaplanes. Flight-Lieutenant C. Boumphrey DFC, was in command and it was to be seven months before he became a squadron-leader.

The Squadron's Fairey IIIDs were developments of a design that had first appeared during World War One. Outwardly, it had changed little from the first of its line and still possessed the flat-bottomed wooden floats of that era, supplemented by smaller floats at the wingtips and a fifth fitted under the tail. The Rolls-Royce Eagle VIII of the prototype and early production aircraft was replaced by a 450 h.p. Napier Lion of exemplary reliability. With a range of under 500 miles, the search area of the Malta-based Fairey IIIDs was not large.

Although handicapped by a lack of any reserve aircraft No. 202 managed to put up five seaplanes for an air display in May 1929 which marked the first public appearance of the new No. 202. Calafrana (the spelling eventually became Kalafrana) was not an ideal place for seaplanes but had been built during World War One as the best spot available. Heavy swells had stopped flying on numerous occasions and the squadron looked forward to newer aircraft capable of operating in sea conditions which put its seaplanes ashore.

It was to be a long wait for flying-boats. No. 202's replacements were allotted during the summer of 1930. During this time, two Fairey IIID accidents occurred. On 5 June 1930, a IIID (S1078) spun into the ground from 3000 feet and killed the crew; and, on 7 July, another IIID (N9730) collided with a dinghy in the bay, killing the occupant. Nine days later, a Fairey IIIF (S1374) was delivered to begin a re-equipment programme.

Despite the fact that they still wore 'sea-boots', the IIIFs were a marked improvement on their ancestors. Their engines were still Lions but of increased power and fitted into a cleaned-up nose. Floats were more streamlined but, unfortunately, were in short supply. This delayed the new seaplanes coming into full service immediately but, by the time No. 202 had been flying IIIFs for a year, the squadron was ready to begin 'showing the flag' around the Mediterranean.

Leaving Malta on 14 July 1931, the squadron visited Augusta in Sicily, Corfu, Athens, Mirabella (Crete), Aboukir near Alexandria and Sollum, to become well-known when war broke out in the Western Desert. The flight's first casualty occurred with almost indecent haste. A Fairey IIIF (S1382) had total engine failure soon after take-off and was towed back to Malta by the sloop *Aberdeen*. Later, another IIIF (S1380) just made it to Corfu with an engine water-cooling leak but the

1 Fairey IIIF Mk.IIIM S1373 served throughout its career with No. 202, being one of the first to arrive at Malta. For foreign cruises, it had the squadron's badge of knight's shield and albatross applied and an identity number painted on the fin. After 1020 hours of flying, S1373 was struck-off-charge as not being worth overhauling, having been replaced at Malta by a Supermarine Scapa.

2 Airco (de Havilland) D.H.4 D8407 joined No. 202 Squadron on 16 August 1918 and survived the war to be transferred to No. 98 Squadron on 15 March 1919.

3 One of No. 202's original equipment inherited from No. 481 Flight in January 1929, Fairey IIID N9571 was taken out of service in 1930 and replaced by a Fairey IIIF.

© Hylton Lacy Publishers Limited

1 Handley Page Hastings Met. 1 TG565 differed from most of its contemporaries by the addition of a radome in the nose. Revised colour schemes for No. 202's aircraft came into effect in 1963.

2 Supermarine Scapa K4196 was operated by No. 202 Squadron between August 1935 and May 1937. After being flown back to England, it was scrapped in October 1937.

3 Consolidated Catalina IVA JX208 was delivered to No. 202 Squadron on 27 March 1944 but on 19 December 1944 crashed at Castlegregory, near Tralee, with the loss of its entire crew of nine.

© Hylton Lacy Publishers Limited

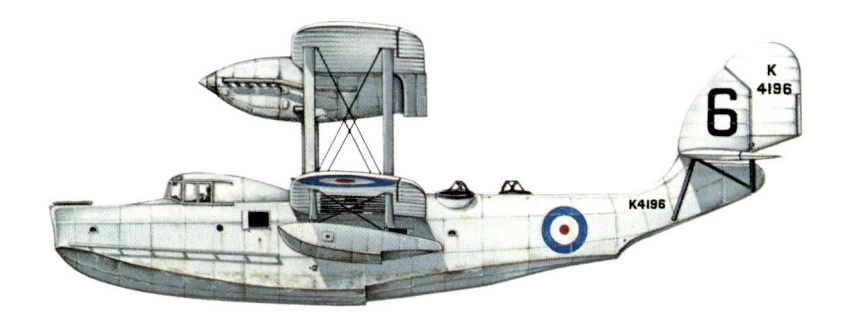

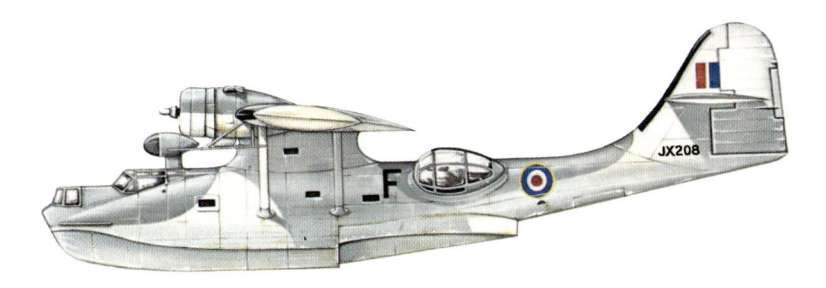

remaining floatplanes completed the cruise with only minor ailments.

An early example of air-sea rescue operations from Malta occurred on 16 February 1932. An Italian airline founded in January 1925, SANA, had begun a regular flying-boat service between Rome and Tripoli with stops at Augusta and Malta. One of their *Wal* (Whale) flying-boats, built under licence in Italy from a Dornier design, sent out an SOS from 50 miles south of Malta. Flying-Officers C. W. Grannum and J. A. Rutherford set out in S1384 to look for it and located the downed aircraft with commendable speed. Circling the area, the seaplane's radio calls directed the destroyer *Brilliant* to the scene and she took the Wal in tow for Malta. Unfortunately, just outside Marsa Scirocco Bay, high wind capsized the flying-boat. HMS *Brilliant* cast-off the tow and brought the crew and passengers ashore. Next day, two more IIIFs (S1380 and S1382) set off to search for the wreck, and located it 15 miles from Malta. It was later towed in to Calafrana and slipped for inspection.

In May, the squadron went to the rescue of the crew of a Fleet Air Arm Blackburn Ripon from the carrier *Glorious* and rescued the pilot by landing alongside, the observer being picked up by a IIIF of 447 Flight.

The squadron set forth on another cruise on 20 June 1932. Four floatplanes left for Khartoum via Sicily, Greece, Cyprus, Lebanon and Egypt on a 25-day trip during which much valuable experience was gained in operating away from base. In 1933, the destination was the Adriatic with visits to Kotor and Split. At the same time, a Supermarine Southampton IV passed through on its way to tropical trials at Port Sudan after a non-stop flight from Gibraltar in 12 hours. Malta's 'Flying Boat' squadron regarded it with envy unaware that it was the first sign of their future equipment.

1 Fairey IIIFs beached at St Paul's Bay, Malta (*Photo: MoD. H842*)

2 No. 202's six Fairey IIIFs lined-up at Kalafrana with a Fleet Air Arm Fairey Flycatcher fighter at the end of the line. Carriers normally fitted several aircraft with floats while they were berthed in Grand Harbour and these were usually based at Kalafrana (*Photo: MoD. H1623*)

3 The Fairey IIID was normally a Fleet Air Arm type operating with a land undercarriage, No. 202 being the only non-naval squadron to fly them (*Photo: MoD. H1341*)

4 Five Fairey IIIFs in echelon formation off the Maltese coast in 1935. No. 202's badge is carried on the forward fuselage in readiness for the squadron's summer cruise and S1385 is flying formation leader's pennants from rudder and wingtips (*Photo: No. 202 Squadron collection*)

5 Five Fairey IIIFs moored on the Nile near Khartoum in June 1934 during a cruise from Malta to the Sudan. At the end of the line is a Fairey Gordon floatplane of No. 47 Squadron which had a seaplane flight for local patrol duties (*Photo: IWM. CM4499*)

SANA's Wals still seemed to play a large part in No. 202's activities. Flying-Officer Dobell was landing their IIIF (S1385) in the Bay on 7 November 1933 when the SANA boat landed at the same time but at 90 degrees to the wind. Damage was fortunately minor when the two met. Seven days later, another SANA boat had to be located in the sea between Malta and Sicily and was duly towed in. In March 1934, Calafrana's small craft were out to the south of Malta to collect yet another Wal spotted by No. 202's Commanding Officer. It was all good reconnaissance practice; as was the 1934 cruise, back again to Khartoum with five floatplanes.

May Day 1935 was a red letter day for the squadron. A Supermarine Scapa (K4192) arrived from England to begin the replacement of No 202's IIIFs. It was the production version of the Southampton IV seen two years earlier and No. 202 (Flying Boat) Squadron at last justified its title. A second Scapa (K4193) arrived on 24 May, to be followed by three others (K4194, K4195 and K4196); the last arriving on 3 August.

The all-metal Scapa was a logical successor to the earlier Southampton which had served the RAF so well for ten years. Two 525 h.p. Rolls-Royce Kestrel IIIs gave it a range of over 1000 miles and a speed of 140 m.p.h. Three gunners still occupied unprotected positions but the pilots had a cabin and there was accommodation in the hull behind the cockpit for the rest of the crew when not manning their machine-guns.

No. 202's boats arrived just in time to become engaged in more active operations. The international crisis caused by the Italian invasion of Ethiopia had more repercussions on British units than those of other countries not directly involved in the war. In support of a League of Nations resolution, many RAF units moved to the Middle East. No. 202 began flying anti-submarine patrols around Malta on 9 October 1935 and aircraft began using other mooring areas in case of air attack. Within a year, most of the squadrons were back home and tension in the Mediterranean relaxed.

Two Scapas (K4194 and K4196) left Malta on 12 December 1936 for a cruise to West Africa and followed the African coastline as far as Freetown where one (K4194) lost an engine. The remaining Scapa continued on to Lagos while an engine change was carried out with the aid of two jetties and a fortunate lack of swell. Both flying-boats returned to Malta on 23 January 1937, having surveyed potential flying boat bases, many of which would be used during World War Two.

During September 1937, Saro Londons began to be flown out from England to replace No. 202's Scapas. The latter were flown back by ferry crews and it was during this exercise that the sole Scapa (K4200) to be lost by No. 202 crashed. While taking off for a full-load test in preparation for its long flight home, this Scapa hit a bad swell and broke-up; fortunately, without injuring any of the crew seriously.

Almost immediately, the Londons and remaining Scapas became involved in anti-submarine patrols. Events in the Spanish Civil War had been complicated by the support given to both sides from outside sources. Germany and Italy sent war material to equip the Fascist forces while the USSR saw the opportunity of installing a Communist government by supplying arms to the Republicans who, if successful, would be taken over by Russian-trained elements. Several neutral merchant ships were torpedoed, apparently by trigger-happy Spanish submarines which were later found to have been Italian. At the Nyon Conference in September 1937, it was agreed that an International patrol would be set up which, in effect, would consist of British and French ships and aircraft.

While British-based Short Singapore flying-boats descended on the French seaplane station at Algiers/Arzeu, No. 202 began patrols from Malta. The squadron joined French seaplanes from Bizerte/Karouba and flying-boats from Berre to form a circle round the western Mediterranean. Each boat carried four 112-lb. bombs and full armament; later amended to two 250-lb. bombs. Twenty-three patrols were flown from 20 September till the end of the month, all but one by Scapas. 50 were added to this total during October and 27 more in November; with Londons playing an increasing part as they replaced Scapas. Up to 20 December, when patrols were suspended, 11 more brought the total to 111. Sinkings ceased, more probably due to Italian diplomatic nervousness than fear of a successful attack. Several destroyers had dropped depth charges and the chance of a submarine being attacked coinciding with the necessity of announcing an 'accident' would have taxed the powers of the Italian Foreign Office considerably in providing a believable explanation.

On 10 December 1937, two more Londons (K9684 and K9685) arrived to complete re-equipment. Nine months later another crisis brought No. 202 to action stations again. In the weeks of the Czechoslovakian crisis, RAF squadrons were placed on a war footing. No. 202 was ordered to move to Alexandria. In consequence, on 26 September 1938, four Londons flew non-stop to Alexandria, followed next day by two more. Ground crews sailed in the depot ship *Maidstone* which became a floating base for the squadron at Alexandria on arrival. On 10 October, she sailed for Malta followed by the squadron's boats as the Munich Agreement gave 'peace in our time'. The RAF was distinctly lacking in confidence in this political dream and during the next few weeks No. 202's Londons photographed ports and airfields in Sicily and Libya, in addition to the restricted island of Pantellaria.

During April 1939, a further move to Alexandria was begun but cancelled before the sea party left. At the same time, the squadron was informed of its impending re-equipment with Short Sunderlands. Two Londons were to be flown to Calshot and on the afternoon of the following day, a Sunderland (N6135) flew in over the moored Londons to join the squadron. Two Londons

(K6932 and K9684) left on 30 April for Calshot and conversion training began on 4 May. Unfortunately, it was not to be. No. 202's conversion was suspended on 10 May and the prized Sunderlands were reallotted to No. 228 Squadron. The pair of Londons returned to Malta three days later.

During June 1939, the squadron managed to fit in a last peace-time cruise. Four boats toured Greece and the Aegean. At the Greek naval station at Phaleron, one of the Londons (K9684) was rammed by a taxiing Greek flying boat. It was, inevitably, No. 202's traditional nemesis—a Dornier Wal.

Time was running out for peaceful activities. A merchant ship, SS *Dumana*, had been chartered to act as a mobile base for No. 86 (General Reconnaissance) Wing, which was to control Nos. 202 and 228 Squadrons. On 27 August, maintenance parties moved into dispersed billets and 'scatter' points manned at St Paul's Bay, Cala Mistra, Komino and the motor torpedo-boat trots in Mgarr Harbour. Kalafrana (which had changed its initial letter to conform with local spelling) remained the main repair base but was vulnerable to attack from the Sicilian airfields only 50 miles away.

When war broke out, No. 202 had six Londons manned by 12 officers and 69 ncos and airmen; 51 short of establishment. Three Sunderlands of No. 228 Squadron were also based on Malta. On 9 September 1939, orders were received for the squadron to move to Gibraltar; and, next day, five Londons were winging their way to the Rock.

Guarding the Straits

The squadron's new home was unique even by the elastic standards of the RAF's ideas of normal bases. The great outcrop of rock projected from Spain on a peninsula of only a few square miles. Honeycombed with caves used as magazines and storage areas, Gibraltar was extremely short of level ground. The town huddled around the base of the rock and the only flat area was the land between the North Front of the rock and the Spanish frontier. Here was to be found the race-course which could serve as a primitive airfield but the proximity of the border and the definition of part of the flat area as 'neutral ground' seemed to prevent its development to any great degree.

A naval base of Gibraltar's importance could not be left without air support. As the key to the entrance to the Mediterranean, its approaches had to be patrolled and No. 202's Londons were now on hand for this task. Operating from Algeciras Bay and hemmed in by neutral territory, flying was difficult even in good weather. Fascist Spain owed its existence to Germany and Italy and was not disposed to be accommodating to

1 Farewell to the Fairey IIIF. The last of No. 202's floatplanes at Kalafrana in June 1935 (*Photo: No. 202 Squadron collection*)
2 No. 202's first Scapa (K4192) taxies into Marsaxlokk Bay on 1 May 1935 with Flight Lieutenant C. W. Dickson nonchalantly manning the side (*Photo: No. 202 Squadron collection*)
3 S1373 flying near the Kalafrana oil installation is fitted with early-type exhaust stubs, later replaced by 'ram's horns'. (*Photo: No. 202 Squadron collection*)
4 Resting on the shingle of Marsaxlokk Bay a Scapa (K4196) beside a Short Rangoon of No. 210 Squadron in 1936 (*Photo: No. 202 Squadron collection*)
5 A London (K9682) just after launching at Kalafrana. In the background are two Supermarine Walrus amphibians of the Royal Navy's Mediterranean Fleet whose aircraft were frequent visitors to the seaplane station (*Photo: No. 202 Squadron collection*)

the occupiers of a piece of territory which it hoped to acquire in the event of a German victory. Straying aircraft were met by anti-aircraft fire from the mainland.

A London (K9683) set out on the squadron's first patrol from Gibraltar on 11 September. At first, a major requirement was the location of any German shipping in the area. Many merchant ships were identified in Spanish harbours and a watch was kept for any excursions. For short-range patrols, the Londons were supplemented by Fairey Swordfish floatplanes from No. 3 AACU (Anti-Aircraft Co-operation Unit). For night landings, the Admiralty tug *St Day* acted as a replacement for a flarepath, using marker lights and a 10in searchlight. In December *Dumana* arrived to serve as headquarters for No. 86 Wing and doubled as an officers' mess.

On 26 December, a No. 202 London (K6931) sighted the German freighter *Glucksberg* off Cadiz and signalled the ship to stop while informing by radio the destroyer *Wishart*. The freighter chose to ignore the London's signal and headed for the mouth of the Guadalquiver as *Wishart* raced to intercept. Cut off from her objective, *Glucksberg* ran herself aground on the Spanish coast and was abandoned.

Throughout the winter, patrols were kept up in spite of a lack of spare parts. Strength was nominally eight boats but the number available each day averaged about three. Maintenance was still being carried out at Kalafrana.

By June 1940 the events in Europe brought about major changes in Gibraltar's tactical position. France retired from the war and Italy joined in for what Mussolini thought were the last few weeks of fighting. At a stroke, Gibraltar was isolated. Instead of friendly bases in Morocco and Algeria to divert to if the weather closed in at Gibraltar, there was now no alternative base closer than Cornwall, Malta or Gambia; none of them were even remotely within reach.

A refusal by the new Vichy regime in France to order the French fleet to a British or neutral port to prevent it falling into German hands resulted in a No. 202 London (K5913) being sent to Oran to photograph the naval base. Two battle-cruisers, two old battleships, a seaplane carrier, six destroyers and four submarines were plotted. The purpose of this flight became obvious on 3 July when, after an ultimatum requiring the French Fleet to place itself out of reach of the Germans was rejected, a force of British ships bombarded Oran and blew up one battleship and damaged many other ships.

Two Sunderlands of No. 228 arrived to supplement No. 202's patrols which had now to cover areas previously the responsibility of French aircraft from North Africa. Being near Italian air bases, Kalafrana had become too vulnerable to permit maintenance to continue there and the Londons were now flown back to England for overhaul. Occasionally, submarines were sighted and attacked and air raids from Italian and French aircraft took place. These caused no damage to the squadron but mechanical failure did. A London (K5260) made a forced landing 20 miles out to sea on 12 August and sank under tow. Another (K5909) en route from Britain failed to arrive and a London set off to search for the missing boat. The crew were fortunate in being located in the open sea by a Royal Navy Blackburn Skua and the London was towed into Gibralter by the destroyer *Forester*.

An attempt to capture Dakar by the Free French during September required the Londons to keep watch on Casablanca to report any movement of French warships. As Vichy fighters were based there, it was not a task which should have been allotted to obsolete flying-boats; but there was no other unit within range. On 13 September, two Londons (K6930 and K5908) took turns to watch the French naval base and were fired on. Next day, five more Londons maintained patrols and one (K9682), with Flight-Lieutenant MacCallum and crew, sent out an SOS and disappeared. Casablanca later signalled that the crew had reached the coast safely. Another London (L7043) was shadowed by four French fighters for a time but was not attacked. In many French units, feelings differed among their personnel, some supporting the Vichy regime and others still hoping for a British victory. A number succeeded in escaping with their aircraft to join the Free French, others avoided action with British aircraft while some were fanatically hostile—they blamed Britain for France's defeat.

On 18 October 1940, hours of fruitless patrolling

paid off. Flight-Lieutenant N. F. Eagleton and his crew (in K5909) and Flight-Lieutenant Hatfield and crew (K5913) sighted oil bubbles and dropped depth charges on the spot which damaged the Italian submarine *Durbo* sufficiently for it to surface and be captured by destroyers. It was to be the London's only success against an enemy submarine.

The Swordfish of No. 3 AACU were transferred to No. 202 Squadron on 27 October 1940 and became 'B' Flight. The three floatplanes were used for patrols over the approaches to Gibraltar and gave good service. Being designed to operate from the decks of aircraft carriers, Swordfish could stand up to rough handling with minimal maintenance.

Wing-Commander T. Q. Horner was subjected to the attentions of a pair of Vichy French Curtiss Hawk 75As while patrolling off Casablanca (in K5909) on 28 January 1941 but escaped with a few bullet holes. Next month, another London experienced difficulties and force-landed in heavy seas while escorting a convoy. The destroyer *Isis* rescued Flight Lieutenant Nicholls and his crew but the flying-boat (K5263) foundered under tow three hours later. A strong easterly gale on 11 April resulted in a No. 202 boat (K6930) becoming airborne while attached to its buoy. On falling back to the water, it turned turtle and the two occupants escaped through hatches and were picked up with difficulty. The Bessoneau hangar housing the floatplanes on Gun Wharf was blown down and a Swordfish (L9770) was wrecked.

These losses were made up on 24 April when two Consolidated Catalinas (AH537 and AH538) were taken over by No. 202. The Catalina was the US Navy patrol boat equivalent of the Sunderland but had only two engines. Earlier versions had set up several records for long-range flights with the US Navy but had not been selected by the British Purchasing Commission as they were not up to the standards of RAF requirements. The improved PBY-5 version lacked armament, its four hand-operated machine-guns offering only light defence against enemy fighters. Despite this, the need for more maritime reconnaissance aircraft resulted in orders being placed after war broke out. Once in service, they proved their worth in very long patrols over areas out of range of enemy aircraft, leaving the heavily-armed Sunderlands to cope with more dangerous regions.

No. 202's new boats were powered by 1200 h.p. Pratt & Whitney R-1830 'Twin Wasp' radials which gave them a speed of 190 m.p.h. and a range of about 4,000 miles. Fifteen-hour patrols were common and some lasted up to 24. Visibility from a pair of large transparency 'blisters' was excellent. During May, three Catalinas arrived at Gibraltar, W8410 being the first to take up station on 7 May.

When, on 4 June 1941, Flight Lieutenant Levien landed, he had completed the last patrol by a London (K5911) of No. 202 Squadron. Three days later, one of the Catalinas (W8407) was lost en route from England; having sent out an SOS warning Gibraltar that its elevator was unusable and that an emergency landing in the Strait would be necessary. A Swordfish (K8354) went out on search but soon the Catalina was sighted over the Bay. Lacking elevator control, W8407 crashed on landing; and, although rescue craft reached the scene immediately, only seven of the crew of nine were rescued.

---

1 'B/202' (K6932) moored in the harbour at Gibraltar. Fitted with the then-standard long-range saddle tank, this London carries four anti-submarine bombs (*Photo: No. 202 Squadron collection*)

2 Gun Wharf, Gibraltar, home of No. 202 from September 1939. The hangar was used to house the Swordfish floatplanes and aircraft were hoisted on to the quay by means of the steam crane whose smoke partly obscures a London (*Photo: No. 202 Squadron collection*)

3 'H/202' is hoisted on to Gun Wharf, Gibraltar. The three-blade propellers have replaced the original four-blade variety (*Photo: No. 202 Squadron collection*)

4 A Catalina flies over Gibraltar in 1942. On the extreme left is the airfield where a Vickers-Armstrongs Wellington is taking-off and Lockheed Hudsons are parked on the edge of the 'Neutral Zone.' At the eastern end of the airfield are parked the Gloster Gladiators of the local meteorological flight and some fighters. (*Photo: IWM. CM 6239*)

Patrols over the convoy routes continued with occasional encounters with enemy aircraft from the west coast of France, usually Focke-Wulf FW 200 Condors. On 25 October, 'A/202' (AH538) with Squadron-Leader Eagleton and crew aboard was searching ahead of convoy HG75 when a submarine was sighted. Running in, the Catalina blazed away with machine-gun fire and dropped two depth charges. Both failed to explode so an escort vessel was signalled. The destroyer Lamerton opened fire on the submarine which was then scuttled—its survivors identified the victim as the Italian Ferraris.

Flight-Lieutenant Powell and crew in 'C/202' (AJ162) caught a U-boat submerging on 2 May 1942 and dropped all depth charges in a salvo on the vanishing craft. This submarine, U-74, was finally despatched by a pair of Royal Navy destroyers. Another Catalina ('F/202'; AJ158) was less fortunate while escorting a convoy bound for Malta on 18 May. Investigating two small ships on fire, the Catalina was attacked by Vichy French Dewoitine D 520 fighters off Algiers and forced down. A destroyer, HMS Isis, rescued the crew, including the badly-wounded pilot. But a Royal Navy Fairey Fulmar carrierborne fighter which came to the scene was immediately shot down by one of the D 520s. A Sunderland of No. 10 Squadron, Royal Australian Air Force, had intercepted signals from the Catalina and arrived overhead to cover the destroyer. Two French fighters were seen approaching and the Sunderland climbed to intercept them. After firing a burst of cannon from long range, the D 520s made off before coming within effective range of the Sunderland's own guns. That evening, Catalina 'C/202' (AJ162) landed near the destroyer to take off the wounded pilot and fly him back to Gibraltar.

To supplement No. 202's Catalinas, some Sunderlands were added to the squadron's strength, T9084 making the first patrol on 26 December 1941. Eighteen served with the squadron at various times during the next nine months. A Sunderland ('R/202'; W6002), with Flying-Officer Walshe and crew aboard, sank the Italian submarine Alabastro on 14 September 1942 with six depth charges in the eastern Mediterranean. Six days later, the last Sunderland patrol was flown and the three remaining aircraft flown back to Britain to be reallocated to No. 119 Squadron.

November 1942 was a busy month for No. 202. Operation 'Torch' was under way and large convoys from Britain and the USA headed for North Africa. By this time, the Gibraltar airfield site on the Spanish frontier had been expanded by building a runway and parking areas. Since the neck of land was too narrow to contain a runway aligned in the only possible direction, a wide tongue of reclaimed land was built out into the bay. As invasion time approached, the 'neutral area' became distinctly warlike as numerous Supermarine Spitfires were prepared for the flight to airfields in Algeria as soon as they were captured. Lockheed Hudsons arrived to augment No. 202's patrols and every inch of ground was used to accommodate aircraft. Now uncertain about an ultimate German victory, and relieved that they had turned down an earlier German suggestion that they should invade the Rock, Spain made only muted protests.

Follow-up convoys increased the concentration of shipping using the Strait and the short duration of French resistance was fortunate as it enabled Gibraltar airfield to be cleared of fighters and allowed patrol aircraft to begin operating from Algeria, supplemented by US Navy amphibians from Morocco. For No. 202, this meant that at last a diversionary base was available from which to escape the weather, if necessary.

German U-boats made many attempts to cut the convoy routes without success. In the process, U-620 was caught by Flight-Lieutenant H. R. Sheardown and crew aboard 'J/202' on 13 February 1943—and sunk while lying in wait for convoy KMS 9. As more anti-submarine aircraft became available and more effective tactics and equipment were evolved, the U-boats changed from hunters to hunted. Sightings decreased as U-boats were forced into operating as far from land as possible.

It was a year before No. 202 added another U-boat (U-761) to its score. Flight-Lieutenant J. Finch and crew (in 'G/202') were on patrol when the radio-operator intercepted a sighting report. Altering course, the Catalina came across two US Navy PBY-5 Catalinas and a Lockheed Vega PV-1 Ventura circling a pair of destroyers. Nearby, the U-boat surfaced, the ships opened fire and a Catalina dropped sticks of bombs nearby. The U-boat then headed south on the surface and 'G/202' began to run up from starboard. A salvo of depth charges straddled the

U-boat which blew up. The assessment board allotted a share in sinking *U-761* to 'G/202' and US ships and aircraft.

On 3 September 1944, No. 202 was ordered to the UK and its spell of duty at Gibraltar was over. On the 5th, the first three Catalinas left for home.

Irish Interlude

No. 202's new base was Castle Archdale on the shores of Lough Erne in Ulster. Sunderlands and Catalinas used the calm waters of the Lough as their base for long patrols over the North Atlantic. Unfortunately, the neutral territory of Southern Ireland lay between them and the sea, necessitating a detour round the mountainous region to the west—this hazard claimed a number of aircraft in bad weather.

No. 202's sea party arrived at Castle Archdale on 19 September, where training began on new radar and with Leigh Lights, an airborne searchlight. Strength of the squadron was now 17 aircraft manned by 197 aircrew. Total personnel was 62 officers and 271 other ranks. Operations began on 16 September.

The weather was very bad during the winter of 1944-45. Two boats were lost through crashing into high ground. One Catalina IVA ('P/202'; JX242) failed to return from patrol on 20 November and was found crashed on Magho Hill on the west side of Lough Erne. Eight of the crew of 10 had perished. Exactly one month later, another Catalina IVA ('F/202'; JX208) crashed into mountains at Castle Gregory, near Tralee with the loss of its entire crew of nine.

Despite their long range, the Catalinas were used mainly for coastal patrols. U-boats had been fitted with the '*Schnorkel*' device enabling them to recharge their batteries without surfacing. Only a tell-tale ribbon of smoke from the diesels revealed the position of the submarine whose activities were now concentrated in coastal waters. The weather finally stopped all activity on Lough Erne by freezing it solid—an unprecedented event. As the ice grew to six inches thick, as many boats as possible were beached while the marine craft section maintained a constant patrol in their refuellers and bomb scows in an attempt to prevent the remaining boats being nipped in the ice. For seven days, Castle Archdale was out of action until a thaw set in.

For the rest of the war, coastal patrols brought no results despite the use of sonobuoys to locate submerged U-boats. Germany's 'Third Reich' crumbled and the war came to an end with the act of unconditional surrender signed at Rheims on 7 May. Two days later, a No. 202 Catalina sighted *U-1058* flying a black flag and heading for Londonderry; the first of a stream of

1 Catalinas of No. 202 Squadron moored in Gibraltar harbour. The struttery fitted to the hull of the Catalina in the foreground carried a radar array; more compact fittings were normal (*Photo: IWM. CM2308*)

2 This Catalina (AH562) has acquired No. 413 Squadron's code 'AX' but no individual letter when photographed at moorings in Gibraltar harbour. Several squadrons supplied aircraft for attachment to No. 202 at Gibraltar (*Photo: IWM. CM2307*)

3 Stores ready for loading on a Catalina for a patrol include bombs, ammunition, flares, emergency kits, rations, parachutes and a dinghy (*Photo: IWM. CM2311*)

4 Catalina IVBs of No. 202 equipped with Leigh Lights fly over Killadeas on Lough Erne. Sunderlands and Catalinas of No. 131 Operational Training Unit are waterborne on the lough (*Photo: No. 202 Squadron collection*)

surrendering U-boats. Patrols were maintained until 2 June—the last being flown by 'S/202'.

With squadrons being disbanded, Sunderlands were now available for the remaining squadrons and Catalina units soon disbanded. On 24 June, Wing-Commander Lindsay made the last flight of a No. 202 Squadron Catalina when he flew over London on the occasion of the stand-down on the Royal Observer Corps. The squadron had officially disbanded on 12 June, 24 crews being posted to No. 220 Squadron to fly Consolidated Liberators in RAF Transport Command.

The Weathermen

In the spate of renumbering that afflicted the RAF squadrons in the years following the end of the war, many famous numbers vanished temporarily. To recreate the pre-war list, many squadrons which still existed—but had very short histories—were renumbered. One of these was No. 518 Squadron at Aldergrove.

Formed on 9 July 1943 as a meteorological unit, No. 518 had no hope of remaining under its original squadron identity. On 1 October 1946, it was renumbered No. 202 Squadron. Its equipment consisted of converted Handley Page Halifax bombers fitted out with thermometers and barometers for meteorological flights.

British weather depends mainly on a westerly wind and what happens over the Atlantic usually affects the British Isles within a short period. During the war, the fate of thousands of aircraft operating over Europe depended on good weather forecasting as unexpected meteorological changes could mean that many would be lost returning to bases which had been closed by fog, rain or low cloud. Several squadrons were formed to maintain regular meteorological flights not only over the sea but also over enemy territory using Supermarine Spitfires and de Havilland Mosquitos.

The squadron's wartime Halifaxes were far from ideal for long over-water flights in all weathers. Irregular meteorological flights were of little use to the weather experts at the Air Ministry. An international flotilla of corvettes and frigates was positioned in the Atlantic at fixed points to report weather conditions but only aircraft could verify high-altitude data. Many airlines had opened transatlantic routes soon after the end of the war and the Lockheed Constellations, Boeing Stratocruisers and Douglas DC-4s and DC-6s of the time were very dependent on good weather reporting, being relatively short-ranged aircraft by comparison with the jet airliners that replaced them.

In 1950, a replacement for the old Halifaxes became available. The Handley Page Hastings was the RAF's first post-war transport and was a roomy four-engined aircraft which could provide ample space for meteorological crews and their equipment. Its well-tried Bristol Hercules sleeve-valve air-cooled radials were reliable and gave the Hastings a cruising speed of nearly 300 m.p.h. when required with a useful range to enable it to carry out all the tasks allotted to the squadron.

For many years, a Hastings heading out over the Atlantic was a familiar sight in Ulster. When finally they ceased their sorties, it was because their role was about to be overtaken by a device unthought of when No. 202 first set off with their thermometers. Orbiting in space around the world was a meteorological satellite sending back pictures of the weather far below. In such an environment, Hastings were as obsolete as rowing galleys and the squadron was disbanded on 31 July 1964.

Search and Rescue

By 1964, the all-yellow Westland Whirlwind helicopters of the RAF had become familiar sights along the beaches of the British Isles. Two squadrons maintained detachments at stations along the coast for the purpose of rescuing RAF crews unfortunate enough to end up in the sea. In addition, the helicopters were frequently called upon to rescue holidaymakers stranded by the tide, yachtsmen caught by squalls, merchant seamen from ships gone aground in inaccessible places and numerous other human beings who had managed to place themselves in the most unlikely predicaments.

No. 228 Squadron at Leconfield in Yorkshire was dispersed around the East Coast when it was informed that as from 1 September 1964, it would be renumbered 202 Squadron. During the previous year, 393 calls had been made on the rescue helicopters, only a very small proportion being in answer to military requests.

In the succeeding years, No. 202 operated from Coltishall in Norfolk, Acklington in Northumberland and Leuchars in Fife, in addition to its Yorkshire-based headquarters at Leconfield. Rescues were made not only from the sea and coastline but also from mountains where aircraft had crashed or climbers had become lost or injured.

When rescues were called for on account of the weather, it meant that the squadron's Whirlwinds were operating in marginal conditions. One especially hazardous rescue was when the oil rig *Sea Gem* collapsed in heavy seas 40 miles off the coast of Lincolnshire on 27 December 1965. In gale force winds and the temperature below zero, a Whirlwind from Leconfield arrived over the thick patch of oil that marked the place where *Sea Gem* had stood. Three survivors were seen in the water and the helicopter winched down a member of its crew, Sergeant J. Reeson, into 20 foot high waves. One man was successfully hooked onto the cable and hauled up to safety; then to be lowered onto the heaving deck of a ship standing by. Although the rescue crewman had sustained injuries from hitting the mast of the ship and had been affected by seawater and oil swallowed during the first rescue, a second attempt was made. This resulted in success and another man was winched down onto the deck of the rescue ship.

By now, their rescuer was in a state of collapse and had to be replaced by the navigator who managed to collect a third survivor from the sea. For determination and courage, Sergeant John Reeson of No. 202 Squadron was awarded the George Medal.

Today, the Whirlwinds still clatter out from their bases to the rescue of those in peril. Their effective pick-up range is only 82 miles and replacement with something larger and possessing longer-range capability is overdue. But no matter how elderly their helicopters may be, No. 202's Whirlwind crews are always standing by to answer any call for help from those in danger on land or on the sea.

1 Hastings 'B/202' (TG623) heads out from Aldergrove on a *Bismuth* patrol over the North Atlantic to gather meteorological information (*Photo: MoD. PRB24284*)

2 A Whirlwind of No. 202 brings help to a snowbound farm in Yorkshire cut off from the outside world by blizzards. (*Photo: Yorkshire Post*)

3 To spotlight enemy U-boats at night, Coastal Command aircraft used the Leigh Light. On a Catalina, this was fixed in a self-contained pack under the starboard wing (*Photo: Saunders-Roe 220*)

SQUADRON BASES

Bergues, France	1 April 1918
Varssenaere, Belgium	18 November 1918
Driffield, Yorkshire	24 March 1919 to 22 January 1920
Alexandria, Egypt	9 April 1920 to 16 May 1921
Calafrana, Malta	1 January 1929
Ras-el-Tin, Alexandria, Egypt	26 September 1938
Calafrana, Malta	13 October 1938
Gibraltar	10 September 1939
Castle Archdale, Fermanagh, Ulster	19 September 1944 to 12 June 1945
Aldergrove, Antrim, Ulster	1 October 1946 to 31 July 1964
Leconfield, Yorkshire	1 September 1964

Detached Flights as follows:
'A' Flight : Acklington, Northumberland
'B' Flight : Leconfield, Yorkshire
'C' Flight : Leuchars, Fifeshire
'D' Flight : Coltishall, Norfolk

SQUADRON EQUIPMENT
Period of Use and Typical Serial and Code Letters

Aircraft	Period	Serial/Code
Airco D.H.4	April 1918 to March 1919	A7986
Airco D.H.9	May 1918 to September 1918	B7630
Short 184	April 1920 to May 1921	
Fairey IIID	January 1929 to September 1930	S1078
Fairey IIIF	July 1930 to August 1935	S1386 (5)
Supermarine Scapa	May 1935 to December 1937	K4191
Saro London II	September 1937 to June 1941	K5264 (TQ-L)
Fairey Swordfish I	September 1940 to June 1941	K8354 (TQ-D)
Consolidated Catalina IB	April 1941 to January 1945	AJ160 (TQ-S)
Consolidated Catalina IV	October 1944 to June 1945	JX242 (TQ-P)
Short Sunderland I, II	December 1941 to September 1942	W4024 (TQ-N)
Short Sunderland III	March 1942 to September 1942	DV962
Handley Page Halifax Met.6	October 1946 to May 1951	RG780 (TQ-J)
Handley Page Halifax A.9	August 1949 to December 1950	RT786 (TQ-A)
Handley Page Hastings Met.1	October 1950* to July 1964	TG565 (H)
Westland Whirlwind HAR.10	September 1964 to date	XP403

*Conversion training began in July 1950 before allocation of squadron aircraft

COMMANDING OFFICERS

Sqn Cdr B. S. Wemp	1 April 1918
Major R. Gow	7 April 1918
Capt J. Robinson	1 January 1920
S/Ldr W. G. Sitwell	21 March 1920
S/Ldr P. A. Shepherd	15 September 1920
S/Ldr C. Boumphrey DFC	1 January 1929 *
S/Ldr R. H. Kershaw	28 January 1930 †
S/Ldr H. W. Evens	31 July 1931
S/Ldr A. H. Wann	7 September 1932
S/Ldr J. H. O. Jones	17 May 1934 ‡
W/Cdr E. F. Turner AFC	7 December 1935
W/Cdr G. W. Bentley DFC	19 March 1938
W/Cdr E. A. Blake MM	29 March 1939
W/Cdr A. D. Rogers AFC	15 March 1940
S/Ldr T. Q. Horner	28 August 1940
W/Cdr L. F. Brown	4 March 1941
W/Cdr A. A. Case	2 February 1942
W/Cdr B. E. Dobb	6 January 1943
W/Cdr G. P. Harger DFC	12 December 1943
W/Cdr D. S. Lindsay DFC	7 November 1944
W/Cdr L. Coulson	1 October 1946
W/Cdr J. R. Armistead DFC	10 April 1947
S/Ldr E. W. Deacon DSO DFC	10 March 1948
S/Ldr T. A. Cox DSO DFC	10 November 1949
S/Ldr F. Ellison AFC	10 January 1951
S/Ldr G. T. Thain DFC	3 March 1953
S/Ldr R. Wood DFC	21 March 1955
S/Ldr C. A. Sullings AFC	19 March 1957
S/Ldr K. J. Barrett	17 March 1959
S/Ldr M. J. Davis	9 March 1961
S/Ldr C. J. Petheram	17 March 1962
S/Ldr B. A. Spry	2 March 1964
S/Ldr G. Stafford	1 September 1964
S/Ldr D. E. Brett	19 October 1965
S/Ldr W. A. Gayer	29 May 1967
S/Ldr K. Henry	7 July 1967
S/Ldr P. T. Taylor AFC	9 July 1969
S/Ldr R. G. Reekie	8 August 1971

*Flt Lt until 17 July 1929
†To W/Cdr 1 January 1931   ‡To W/Cdr 4 June 1934

204 Squadron

1 A Southampton I (N9900) shows its five open cockpits as it flies over the English Channel (*Photo: MoD. H657*)

2 A Scapa (K4191) refuelling from the chartered tender *Pass of Balmaha* in Alexandria harbour. In the background, HMS *Barham* (*Photo: MoD H331*)

3 A reason why the nose gunner of a Southampton was in no position to help to repel stern attacks is obvious from this photograph of a No. 204 boat under attack by a Bulldog (*Photo: Flight Int'l. 14807*)

4 A Bristol Bulldog of No. 54 Squadron carries out practice fighter attacks on a No. 204 Squadron Southampton over Plymouth Sound (*Photo: Flight Int'l. 14805*)

Biplane Days

Like so many of the former Royal Naval Air Service squadrons, No. 204 Squadron first came into being in northern France. As No. 4 Squadron, RNAS, it had been formed at Dover on 25 March 1915, becoming No. 4 Wing when the Royal Navy changed the nomenclature of its air units. On the last day of 1916, No. 4 became a squadron again while based at Coudekerque, just outside Dunkerque. During March 1917, it replaced its two-seater Sopwith 1½ Strutters with Sopwith Pup single-seat fighters and three months later received the agile Sopwith Camel.

With the formation of the Royal Air Force, No. 4 Squadron, RNAS, became No. 204 Squadron, RAF but was destined to retain a maritime connection during most of its subsequent history. For the remaining months of World War One, it operated over the Belgian coast, covering naval craft and intercepting enemy seaplanes engaged in spotting for the coastal defence batteries. When, in October 1918, the German army began falling back from its lines in Belgium, No. 204 began ground-attack missions over the retreating enemy. By the end of the war, the Belgian coast was clear and the squadron had moved eastwards to Heule near Courtrai. There it remained until February 1919, when the squadron personnel were posted to Waddington for demobilization. On the last day of 1919, the squadron ceased to exist.

January 1929 had seen the translation of the coastal reconnaissance flights into squadrons and, on 1 February 1929, a new unit was formed to supplement the home-based flying-boat force. It was numbered 204 Squadron and its home station was the old wartime seaplane base at Cattewater in Plymouth Harbour, later to be renamed Mount Batten. The squadron's five Supermarine Southamptons were to be tended by an establishment of 11 officers, one warrant officer, 14 ncos and 60 airmen, commanded by Squadron-Leader F. H. Laurence MC.

Training began with navigation, bombing, direction finding and photography high on the list of priorities for a maritime reconnaissance unit. A series of flights were made to take photographs for the 'Coastal Air Pilot' which was being prepared by the Air Ministry. In July 1930, a summer cruise was made to Dublin and Queenstown, the latter still used as a naval base by agreement with the Irish Free State. August 1931 found four boats cruising up to Oban to cover more of the coast for the 'Coastal Air Pilot.'

During 1932, the squadron went further afield. On 7 August, three Southamptons left for a trip to the Mediterranean with the Prince of Wales. Naples, Split, Venice, Corfu and Malta were all visited in a 19-day cruise. In August 1933, the destination was the Baltic, four Southamptons leaving Felixstowe on 28 August for Esbjerg in Denmark but only two arriving. One (S1647) force-landed at the Dutch naval base of Den Helder and the other (N9900) stood-by for a while before flying on direct to Copenhagen. After repairs, S1647 joined the other three boats at Stockholm. After a short stay in Finland, the squadron returned home by way of the Baltic states; N9900 breaking adrift at Reval (Tallinn, Estonia) and being damaged. The remaining boats arrived back at Felixstowe on 22 September while N9900 reached Mount Batten after repairs during October.

The elderly Southamptons kept flying but had their share of engine problems common to many aircraft of their generation. On 23 May 1935, No. 204 finally lost a boat through engine failure. The squadron was flying patrols from Felixstowe during exercises with the Home Fleet when Flight-Lieutenant L. G. Martin (in S1123) picked up an SOS message from Squadron-Leader A. W. Fletcher, the Commanding Officer, in S1058. In gale force winds and over high seas, S1123 located S1058 heading for the Humber with smoke and steam trailing behind the starboard engine. Half-an-hour later, the two boats arrived over the Fleet and S1038 landed with a seized engine. The destroyer *Crusader* took the flying-boat in tow but heavy seas broke up the aircraft and the wreck was scuttled.

On 23 August 1935, a Supermarine Scapa (K4197) arrived as the first of No. 204's replacement aircraft. Three more arrived during September, including the prototype S1648. On 23 September, three Scapa left for Egypt followed two days later by the fourth. Italy's invasion of Ethiopia had caused a deployment of British units to the Middle East and No. 204's new base was at Aboukir, near Alexandria. The prototype Scapa (S1648) force-landed at Bizerte in Tunisia and, after repairs, left for Malta but again had to put down on the sea and was towed into Kalafrana. Another Scapa (K4192) was despatched from England to replace it and was joined by K4199 to bring strength up to five.

When the depot ship *Manela* arrived at Alexandria in October, the squadron moved to take up its base aboard. For the next nine months, No. 204 remained in the Eastern Mediterranean before being ordered home. On 27 July 1936, the first two left for Mount Batten via Malta, Algiers and Lisbon. While passing Gibraltar at 3000 feet, a Spanish Government destroyer fired on them, hitting K4198 in the tailplane, probably under the impression that the two aircraft were among those transporting Franco's troops from Morocco to the civil war in Spain.

On 10 October 1936, a Saro London (K5263) arrived as the first of No. 204's new boats and next month three Scapas were transferred to the Seaplane Training Squadron at Calshot. Starting-up Londons proved expensive during the first three months of 1937. On 2 February, K5908 caught fire when starting engines at Portland and both port mainplanes were burnt out. On 31 March, a similar accident resulted in K5910 burning for two hours with considerable damage. The squadron establishment was increased to six Londons on 1 April with two more as immediate reserve aircraft. By June, No. 204 Squadron had received three more boats to fill its strength. On 9 August, four Londons flew to Gibraltar in 11 hours non-stop for exercises in the Mediterranean. This long-range practice was useful for the squadron's next major task.

The 150th Anniversary of Sydney, Australia, was celebrated in 1938 and five Londons left Mount Batten on 2 December 1937 for the long flight. Wing-Commander K. B. Lloyd led the formation (in K6930) with Air Commodore S. J. Goble as passenger; and, after twenty refuelling stops, arrived on schedule at Sydney on 25 January 1938. K6927 was missing, having force-landed in the Bay of Bengal and had been towed into Akyab in Burma by the Indian cargo steamer *Jalagopal*. K9686 flew out from Britain as a replacement but suffered the misfortune of having a propeller disintegrate near Melbourne. There was an obvious weakness in the wooden blades and the squadron remained at Melbourne to await a shipment of replacement propellers. It was not until 4 April that the five Londons left for home, arriving back at Mount Batten on 29 May after a 30,000-mile cruise.

The Munich crisis in the Autumn of 1938 found No. 204 at Stranraer on an armament training course. All boats were recalled to base and only test flights permitted apart from a special patrol to locate the German 'pocket battleship' *Deutschland* reported to be heading out into the Atlantic through the English Channel. The Londons were given code letters but had not been camouflaged before the state of readiness was relaxed. In June 1939, a detachment was sent to Pembroke Dock to take over Short Sunderlands for the squadron. Back at Mount Batten, the Londons were handed over to No. 240 Squadron and all mooring buoys were lifted and relaid at wider intervals to make room for the bigger Sunderlands. By the end of July No. 204 was operating its new boats.

The Flying Porcupines

Mobilization orders came through on 2 September 1939 by which time No. 204 had six Sunderlands available for operations with two more in immediate reserve. Reservists arrived within hours and over the next few days were integrated in the squadron's various sections, the Munich alarm having served to iron out some of the inevitable snags. Plymouth's twinkling lights disappeared as a 'black-out' was enforced and the Sound was full of scurrying escort vessels preparing to gather in the scattered merchant ships in the Western Approaches. Convoys had already been organised and at 0430 on 4 September, 'D/204' (L5799) was airborne on its way to escort a convoy in St George's Channel. Returning after nearly nine hours on patrol, the Sunderland was greeted by a salvo from an anti-aircraft battery which had so far failed to acquire sufficient expertize to recognize even a Sunderland.

Anti-Submarine bombs were dropped for the first time when 'C/204' (N9021) sighted a wake 30 miles south-west of the Lizard on 8 September. On the previous day, a detachment had been sent to Stranraer; while, on 16 September, three Londons from No. 240 Squadron at Invergordon had arrived at Falmouth to relieve No. 204 of convoy duties—enabling the bigger boats to be used for anti-submarine patrols. These were normally of 10 hours duration and were concentrated on specific areas about 200 miles out into the Atlantic.

The first patrol on 18 September opened an eventful day for the squadron by sighting a submarine on the surface 250 miles south-west of Plymouth which dived as 'D/204' approached. After carrying out two attacks, dropping four bombs on each run, a black object was seen to come to the surface momentarily after the second but no further trace was found.

Flight-Lieutenant J. Barrett (in 'E/204'; L5802) was patrolling later in the day when his radio-operator picked up an SOS message from the merchant ship *Kensington Court*. The position was about 100 miles south of the Irish coast and 100 miles away from the Sunderland. He arrived to find the torpedoed ship sinking by the bows and a Sunderland of No. 228 Squadron waterborne nearby. Thirty-four men had been able to leave the ship in a single lifeboat and twenty of them were embarked while 'E/204' carried out a quick search of the area for U-boats before landing to pick up the remaining 14. As an example of the Sunderland's capabilities, the event was given considerable publicity and Barrett was awarded the Distinguished Flying Cross.

During October, the squadron lost two of its boats. On 13 October, Flight Lieutenant E. L. Hyde with a crew of 12 was returning from patrol (in 'J/204'; N9045) when fuel ran out near the Scillies. The Sunderland landed in 50-foot waves in a gale-force wind on only two engines, the other two having been shut down in case all four should cut on final approach. As the big flying-boat splashed into the sea, the port float was carried away. Six of the crew scrambled out onto the starboard wing to balance the aircraft while distress messages went out. Fortunately the Dutch freighter *Bilderdyke* was within sight and circled the stricken flying-boat. Nine of the Sunderland's crew set off in rubber dinghies to meet a lifeboat from the ship while the remainder on the wing lit distress flares. Taking up position to windward, the Dutch skipper allowed his ship to drift downwind to the Sunderland and made contact with a crash. Fortunately, a heaving line was secured to a propeller which prevented the flying-boat turning turtle and the remaining crew members were hauled aboard. As soon as *Bilderdyke* cast off, 'J/204' turned turtle and sank. The Dutch ship was en route for New York but the destroyer *Icarus* intercepted her and took of the RAF men. Calculations later showed that the bad weather had raised the Sunderland's fuel consumption to an exceptional level and tests were carried out to arrive at safe limits for patrols.

Three days later, 'B/204' (N9030) was returning from patrol at night and in the darkness the pilots misjudged their height and crashed into the Sound just outside the breakwater with the loss of four of its crew of 11.

Sunderlands were in short supply and on 1 November, No. 204 had only seven of its establishment of eight boats on strength. One of these had been damaged in a collision with a London in August and was still awaiting parts at Mount Batten. None of the aircraft had gun sights or camera mountings. The weather was often very bad and it was fortunate that the specialized training required for flying-boat pilots had resulted in the 'Flying Boat Union' whose highly-trained members were not often diverted from their nautical environment. Though relatively sophisticated by 1939 standards, the Sunderlands lacked many items which were to be regarded as essential within a short time. Even auto-pilots were lacking and the long patrols had to be flown manually. The auto-pilot tested in January 1940 resulted in differing opinions on its worth. The Royal Aircraft Establishment approved of it. The squadron's pilots suggested that it crept 30 degrees every five minutes and so left Sunderlands in a position of possibly emulating a certain mythical bird which flew in ever-decreasing circles.

April 1940 brought a flurry of activity. On 2 April, four boats were ordered to Sullom Voe in the Shetlands and next day 'F/204' (N9046) took off for a patrol off the Norwegian coast with Flight-Lieutenant Phillips and crew aboard. Four hours after leaving base, two aircraft were seen approaching low over the water and were identified as Junkers Ju 88s. Both circled the flying-boat and carried out attacks at fairly long range before four more Ju 88s appeared on the scene. Anti-aircraft fire from a British convoy broke-up the formation but later all six reappeared and began to make attacks from astern on the Sunderland. Phillips took the flying-boat down to wave-top height to prevent attacks from below and Corporal Lillie in his tail turret, fortunately now equipped with a gunsight, returned the fire, supplemented by the two dorsal beam gunners when they could bring their guns to bear. When the leading aircraft was only 100 yards away, the

---

1 The prototype Scapa (S1648) served with No. 204 in September and October 1935
(*Photo: No. 201 Squadron collection*)

2 Scapas of No. 204 Squadron fly over Alexandria harbour during the Abyssinian crisis. Visible are two County-class heavy cruisers, eight destroyers, an Admiralty tug and a sloop
(*Photo: Charles E. Brown*)

four 0·303-in guns in the rear turret of the Sunderland opened fire and the enemy aircraft winged over and dived into the sea. Pilot-Officer Armitstead, the second pilot, acted as fire controller from the Sunderland's astrodome and saw a second Ju 88 hit as it came in astern. Two more tried to bomb the Sunderland from 1500 ft but were easily evaded. The five surviving Ju 88s made off and radio monitors in Britain reported that one had made a forced landing. Presumably this, and the other Ju 88 shot down, were among the three Ju 88s lost that day by the 5th *Staffel* of *Kriegsgeschwader* 30. The Sunderland returned to base with various bullet holes and leaking fuel from a damaged petrol tank. The *Luftwaffe* regarded Sunderlands with a new respect. The nickname 'Fliegender Stachelschwein' or 'Flying Porcupine' came later; but 'F/204s' encounter was the first of many. Phillips received the Distinguished Flying Cross and Corporal Lillie was awarded the Distinguished Flying Medal.

Flight-Lieutenant E. L. Hyde left Sullom Voe on 8 April (in 'B/204': N9047) to escort a force of the Home Fleet covering a minelaying operation off the Norwegian coast and was detached to search along the coast for enemy ships. Under solid cloud at 800 ft, the 'B/204' flew northwards in heavy rain. Suddenly, ships appeared ahead and the Sunderland was hauled round to circle the force. As heavy *flak* opened up, the Sunderland crew identified a battle-cruiser, two cruisers and two destroyers. Hit in its fuel tanks, 'B/204' passed on its sighting report and managed to reach base safely.

61

1 An early production Sunderland I, L5798 lacks the dorsal turret fitted to later aircraft. In its place were two open gun positions (*Photo: IWM. CH2310*)

2 A London (K6929) undergoes maintenance on the apron at Mount Batten. A saddle tank is fitted above the fuselage to provide extra fuel capacity (*Photo: MoD. H0000*)

3 A No. 204 Squadron Sunderland moored at Bowmore, Islay, with other flying-boats. The leading boat is noteworthy in that it is an armed Short S.23/M built originally as an *Empire*-class airliner for Imperial Airways and adapted to augment the small long-range force of Coastal Command early in World War Two (*Photo: IWM. MH6666*)

4 Maintenance aboard Londons was assisted by built-in attachment points for portable benches and platforms. The photograph shows the fixing points for the beaching wheels and the stowage of the saddle tank (*Photo: MoD. H0000*)

Not so fortunate was 'D/204' (L5799) which failed to return from a similar mission. Though it reported only 30 minutes from base, it failed to arrive and was claimed destroyed off the Shetlands by German aircraft. A pair of Heinkel He 111s of II./K.G./26 failed to return from a raid on Scapa Flow that day and may have been part of the enemy formation involved.

Next day, Wing-Commander Davis and Flight-Lieutenant Hyde (in 'C/204': N9044) took-off for Trondheim to locate German shipping involved in the invasion of Norway. Arriving over the entrance to Trondheimfjord, a German Navy Arado Ar 196 catapult floatplane was sighted and chased. It turned in behind the Sunderland and opened fire from astern. Unfortunately, at this juncture the door of the Sunderland's rear turret came open and jammed it. After coming under fire from other guns, the Arado flew off to be replaced by another enemy aircraft. Though identified as a Blohm & Voss Ha 140, it was presumably a Heinkel He 115 two-motor floatplane as the former, though similarly two-motor and appearing widely in British aircraft recognition publications, did not go into production. This time the hapless rear gunner found his turret without hydraulic power and despite all effort failed to get it back into action. Bullets from the enemy seaplane perforated two petrol tanks and it made off trailing smoke from the port engine. Hyde climbed into the wing tunnels to plug leaks with plasticine and the Sunderland headed for home.

As most of Norway fell into enemy hands, Sunderlands took parties of troops and survey teams in attempts to gain toeholds in areas dominated by enemy aircraft based on existing airfields. In contrast, the only RAF fighters in the southern half of Norway had to operate from a frozen lake while in the far north, an airstrip was carved out for a handful of Hurricanes and Gladiators. As the German armies rampaged through France, northern Norway became an expendable outpost and Allied troops were withdrawn.

The battle-cruiser *Scharnhorst* had been at Trondheim for some time and was expected to make for home. On 21 June, she put to sea and headed south with a destroyer escort. Flight-Lieutenant Phillips (in 'A/204': N9028) and Squadron-Leader Thomas ('F/204': N9046) were on shadowing duty during the day. 'A/204' was first off, escorted by three Bristol Blenheims from Sumburgh on the southern tip of the Shetlands. Unfortunately, the escort peeled-off to attack a Dornier Do 18 tandem-Diesel flying-boat and then left for base, leaving the Sunderland on its own. Five Fairey Swordfish torpedo-bombers of the Fleet Air Arm were seen flying eastwards and soon afterwards the battle-cruiser and seven destroyers were sighted and reported. The five Swordfish were seen to make a torpedo attack and one crashed into the sea. A Heinkel He 60, an obsolescent biplane floatplane shadowed the Sunderland in turn but restricted itself to dropping a bomb which landed 50 yards away. Just over an hour later, three Messerschmitt Bf 109s arrived and a fourth joined up before they began making beam attacks on the flying-boat. After 15 minutes, one of the Messerschmitts went down into the sea in flames and 'A/204' made for base with holed fuel tanks and the rear turret out of action. The shadowing was taken over by 'F/204' which kept near cloud cover. Messerschmitt Bf 110 two-motor fighters were sighted but, though they kept watch on the Sunderland, they did not attack before the Sunderland was recalled to base. The battle-cruiser *Scharnhorst* reached Germany after more attacks by Bristol Beauforts and Lockheed Hudsons without further damage. Unknown to the British she had already been torpedoed by the destroyer *Acasta* and was out of action for nearly six months.

For the rest of the year, anti-submarine patrols predominated. The squadron was unfortunate in losing three more Sunderlands, all accidentally. Two were lost in one day on 28 October. 'K/204' (P9620) had force-landed out of fuel after 12 hours in the air and two more of No. 204's Sunderlands were sent out on search. Before they could locate the missing flying boat, it had been found and its crew rescued by HMAS *Australia*. Flying-Officer Armitstead (in 'H/204': N9024) returned to Sullom Voe while Squadron-Leader Cumming (in T9045) headed for Invergordon in very bad weather. Arriving in darkness over the Morayshire coast, the Sunderland made for Invergordon but in the darkness was unsure of its position. With 20 minutes fuel remaining, a landing was made near some lights but the starboard float was smashed on landing. A trawler took the Sunderland in tow but the remaining float was stove in and the crew was taken off. Abandoned, the flying-boat sank five miles off Strathie Point. A Sunderland ('F/204': N9046), which had reached the squadron just before the outbreak of war, caught fire and was destroyed at its moorings.

During the summer of 1940, British forces moved into Iceland to prevent any attempt at German occupation after their take-over in Denmark. RAF Coastal Command patrols from Reykjavik flew south to help cover areas of the North Atlantic and in April 1941, No. 204 was ordered to move there. Between 3 April and 7 April, the squadron's Sunderlands shuttled personnel and stores to the new base and the depot ship *Manela* arrived from the Shetlands to become a floating headquarters. From bleak surroundings, patrols set out for the convoy routes.

On 23 April, another Sunderland (N9023) failed to return from patrol, having crashed into an Icelandic mountain with the loss of its entire crew. Six weeks later, on 9 June, 'B/204' (N9047) had been sent to survey a possible emergency base at Thingvalla. Next day it returned to Reykjavik and refuelled. Watchers aboard *Manela* saw smoke coming from the flying-boat and within three minutes the Sunderland was ablaze and sinking. The occupants were taken off by a launch before 'B/204' sank from sight.

The squadron's stay in Iceland was relatively short. At the end of August, six boats left for Gibraltar and Bathurst for service in West Africa. Earlier, four boats had flown to Pembroke Dock and within a few days six had arrived at Gibraltar to fly patrols for two weeks. The depot ship *Manela* sailed for Britain with ground crews on 17 July. The change from the chilly reaches of Icelandic harbours to the hot sun of Gambia was at first appreciated by the squadron's crews but before long they would be remembering with nostalgia the brisk sunny days in the North.

The new base was a sheltered inlet without the natural hazards of mountains and rocks which had obstructed the squadron's previous bases. The humid climate, however, caused many problems with instruments and radio while there was a perennial shortage of spare parts. The bulge of West Africa was a cross-roads for Allied shipping. Enemy U-boats were active but the presence of the Sunderlands and Hudsons from Gambia, Sierra Leone and Nigeria caused them to become cautious and losses were much smaller than they might have been without air cover.

Wear and tear proved far more dangerous to the Sunderlands than the enemy. On 28 June 1942, 'V/204' (T9041) crashed into the sea while on convoy patrol and the crew spent two days in their dinghies before Hudson 'E/200' sighted them. Contact was then lost for 24 hours until the same aircraft spotted the dinghies again and directed the destroyer *Velox* to their aid. As they were taken aboard, 'H/204' arrived overhead. On 16 August, 'E/204' (T9070) caught fire at its moorings and its depth charges blew up. Next day the first production Sunderland L2158 ('M/204') failed to return from patrol and was lost in the wastes of the South Atlantic.

In August 1942, French West Africa was still hostile and on 2 September 'A/204' was shot at and slightly damaged by Vichy fighters while in transit between Gibraltar and Bathurst. On the 13th, the BOAC flying-boat *Clare* (ex-*Australia*, a Short S30 Empire-class boat, registered G-AFCZ) left for Gibraltar and one and a half hours later sent out an SOS. Bodies and wreckage were located three days later by the BOAC Catalina *Guba*.

The invasion of North Africa in November 1942 resulted in the French colonies in north-west Africa

63

coming over to the Allied side. On 13 January 1943, three flying-boats, one each from Nos. 95, 204 and 270 Squadrons, arrived at Port Etienne in Senegal to extend their patrol line northwards. At Dakar, a French-manned Sunderland squadron was formed and was augmented by another squadron with Vickers-Armstrongs Wellingtons. No. 295 Wing, which controlled operations in West Africa, now had four Sunderland, one Catalina, one Liberator and two Wellington squadrons. The Catalinas, manned by New Zealanders, were replaced by Sunderlands in 1944. The two Wellington squadrons were French and South African-manned.

During 1943, five Sunderlands were lost. 'A/204' (JM669) ditched in transit from Gibraltar on 14 April; 'J/204' (JM680) collided with No. 95 Squadron's 'H/95' on take-off on 31 May and sank; 'P/204' (EJ145) landed in the sea off the Spanish colony of Rio de Oro on 17 July; 'X/204' (JM710) dived into the sea off Bathurst on 22 September; and, finally, 'C/204' (W6079) crashed on landing at Bathurst. In 1944 three more boats were destroyed. 'Q/204' (DV991) lost a float while taking-off on 13 July 1944 and its depth charges exploded, killing nine of the crew. 'E/204' (JM672) crashed soon after take-off at Jui, Sierra Leone, and five aircrew perished, while 'B/204' (EK580) lost four of its crew when it crashed on landing at night at Bathurst.

Throughout all this period, the long monotonous patrols went on. While there were few U-boats operating in the area, any relaxation would have brought the enemy down on the West African convoy routes. Despite the vast number of hours flown by No. 204's Sunderlands, not a single U-boat was credited to the squadron when post-war assessments were made, based on captured enemy records. On the other hand, many ships were still afloat that would have been at the bottom of the sea but for RAF Coastal Command's devoted work.

The RAF in West Africa did not remain in being for long after the end of the war in Europe. On 30 June 1945, No. 204, in common with the other units of No. 295 Wing, was disbanded. Of its eight boats, one was transferred to the French at Dakar, five were flown back to England while 'F/204' (DV959) and 'L/204' (EK582) were scuttled off Cape Sierra Leone as being at the end of their useful life. The French units reverted to the command of *Aéronavale* (French Navy). As the last Sunderland took-off and headed for Gibraltar, only the empty huts and the remains of scrapped Sunderlands were left to show that West Africa had once had a powerful maritime air force of its own.

The Post-War Years

The first few years after the end of World War Two found the RAF reorganizing its squadrons into what was expected to be a permanent peace-time structure. Many squadrons were disbanded and their aircraft scrapped; others were renumbered to bring back into the line squadrons which had built up a tradition between the wars. In a few cases, new squadrons were reformed and given famous numbers. One of these was No. 204.

The large air force based in the eastern Mediterranean melted away soon after the end of the war in Europe. Fighter and reconnaissance squadrons were mainly based around the Suez Canal with outposts in Iraq, Aden and Libya. A series of staging posts was set up along the routes to India and the Far East along which flew RAF aircraft to reinforce the units based there.

No. 204 was reformed on 1 August 1947 at Kabrit, a large airfield in the Canal Zone. Forming part of the Middle East Transport Wing, it was engaged mainly in transport flights between RAF bases as far away as India, Kenya and Malta as well as more short-ranged trips. Backbone of the RAF's medium-range transport force was the Douglas Dakota, large numbers of which had been supplied under Lend-Lease during the war. In the absence of any British-designed replacement in 1945, a certain number had been retained by the RAF after Lend-Lease ended while development of the Vickers Viking into a military transport proceeded.

Dakotas had become famous before the war as the Douglas DC-3 airliner and production was accelerated of stripped-down versions for use as troop-carriers and transports. Large numbers were built and the ubiquitous Dakotas served on every front. With the RAF, their greatest contribution to the war effort had been

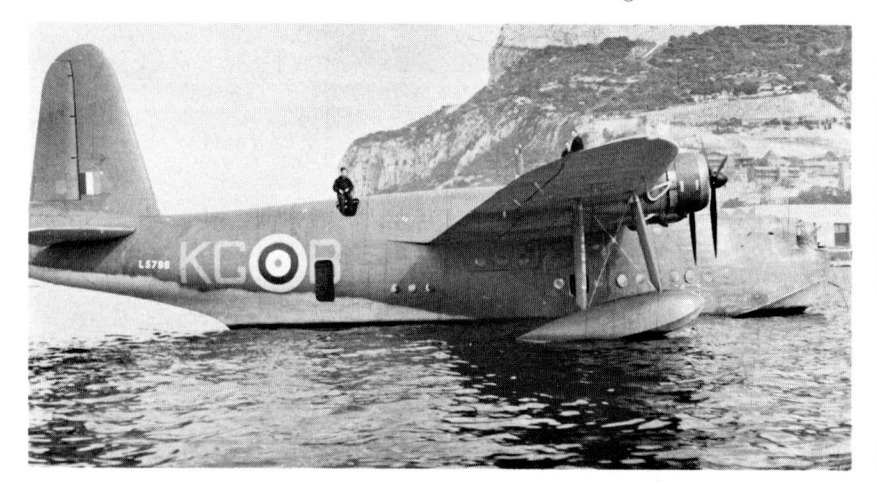

---

1 Moored in Gibraltar harbour, Sunderland I L5798 ('B/204') was used by No. 204 Squadron between July 1941 and September 1943 *(Photo: IWM. CH2309)*

2 Sunderland 'M/204' takes-off from Bathurst. The first production Sunderland (L2158), it joined 204 Squadron in May 1941 and was posted missing on patrol over the South Atlantic on 17 August 1942. The four dorsal masts are ASV aerials *(Photo: IWM. MH 6667)*

3 Sunderland 'F/204' (T9072) undergoes maintenance at its mooring off Bathurst, Gambia *(Photo: IWM. MH6668)*

4 Sunderland III 'E/204' (T9070) served with No. 204 Squadron from December 1940 until it blew up at Bathurst on 16 August 1942 *(Photo: IWM. MH6669)*

to supply the 14th Army in the Burmese jungle in the face of appalling weather and terrain and the massive airlifts to Normandy, Arnhem and across the Rhine.

For two years, No. 204's Dakotas hauled passengers and freight around the area before the last was retired. In May 1949, the military Viking, now designated Valetta C.1, began to replace them. Powered by two 1975 h.p. Bristol Hercules engines, the Valetta could carry a larger load in its more rotund fuselage but its normal operating performance was similar to that of its predecessor. No. 204 was the first of the Middle East squadrons to convert, followed by Nos. 114, 216, 70 and 78 Squadrons in that order. By the middle of 1950, the entire wing had been converted to Valettas though a few Dakotas remained in service for many more years.

By 1951, Egyptian nationalism had gathered sufficient force for the security of the Suez Canal to be threatened. While not anxious to become involved in Egypt's internal politics, the British Government regarded the Canal as a vital waterway and was reluctant to entrust its defence to the highly-incompetent Egyptian Army. A British-manned Canal Zone formed a buffer between the populated part of Egypt and the new state of Israel. Egyptian politicians maintained a barrage of threats against the latter and one result of their oft-repeated promises to exterminate Israel was likely to be the arrival of Israeli forces on the Canal. Britain had few illusions about the abilities of the army and air force it had laboured to train and equip for many years.

Transports of the Middle East Transport Wing flew in over 17,000 reinforcements for the Canal Zone and evacuated the families of the garrison. Agreement was reached on a permanent British garrison for the area based on the Canal, many other installations in Egypt being closed down. No. 204 had moved to nearby Fayid and continued flying its Valettas until 20 February 1953, when the squadron was renumbered 84 Squadron, the latter having had close associations with the Middle East during the inter-war years. Six Valettas were transferred on renumbering, a total of 23 having served with the squadron without serious accident.

Within a year, No. 204 was back in the RAF's order-of-battle. On 1 January 1954, the squadron reformed at Ballykelly, Coastal Command's base in Ulster. On 6 January, its first Avro Shackleton (WB828) was allotted. This was a MR.1 version, the earliest of the Shackleton line, but this mark only supplemented the squadron's main equipment of MR.2s. A pair of MR.1s was used until August 1954 and between April 1958 and February 1960 eight more were issued at various times, the largest number (six) being on hand in November 1958.

The Shackleton MR.2s began to arrive during January and the squadron establishment was set at eight. These long-range patrol aircraft were used for maritime-reconnaissance and air-sea rescue duties, many exercises being carried out in co-operation with NATO naval forces. Detachments were sent on trips which were similar to the pre-war 'show-the-flag' expeditions undertaken by maritime squadrons. One was to South Africa in June 1955 and this helped the South African Government to make up its mind about a replacement for its elderly Sunderlands and an order for eight Shackletons was placed.

The long-expected crisis in the Middle East in the autumn of 1956 found No. 204 back in its old haunts as its Shackletons ferried troops to the area. The following year, it went further afield when the squadron was detached to Australia. Based at the Royal Australian Air Force station at Pearce in Western Australia, the squadron's four Shackletons carried out meteorological observation missions in connection with nuclear tests between August and November 1957.

The arrival of elderly Shackleton MR.1As began in April 1958 to enable the Mark 2s to be progressively modified to Mark 2C standards. This entailed the fitment of equipment to bring the Mark 2's radar and weapon capability up to the same level as the later Shackleton MR.3s. With these updated aircraft, No. 204 remained at Ballykelly until 1971.

During March 1971, the squadron began moving to Honington in Suffolk and became operational there on 1 April. Still equipped with Mark 2Cs, which paradoxically had outlived the later Mark 3s, No. 204's task was shipping reconnaissance and air-sea rescue. All other British-based maritime-reconnaissance squadrons had re-equipped with Hawker Siddeley Nimrods while No. 205 Squadron in Singapore had disbanded at the end of 1971. No. 204 filled the requirements of the Far East with a detachment of Shackletons while others were based at Majunga in the Malagache Republic (or Malagasy; better known by its former name of Madagascar). Until the end of March 1972, Shackletons maintained a watch on vessels using the ports of Mozambique to avoid United Nations' sanctions against Rhodesia.

On 28 April 1972, No. 204 was disbanded and its tasks handed over to the Nimrod squadrons at Kinloss and St. Mawgan. Twenty-three years after its first flight, however, the Shackleton remains in service with the RAF. As the last maritime-reconnaissance Shackletons were withdrawn, No. 8 Squadron at Kinloss were receiving their equipment in the shape of Shackleton AEW.2s with the bulbous early-warning radar scanners. Despite the twenty-year gap since the first Mark 2 took to the air, it appears probable that the distinctive roar of Griffon engines will continue to be heard for many more years.

3·4

1 Short Sunderland I L5798 was the twelfth Sunderland to be delivered to the Royal Air Force and served with Nos. 201 and 210 Squadrons before joining No. 204 on 31 July 1941. After considerable operational flying in West Africa, L5798 was returned to England on 1 October 1943 and was struck-off-charge at Calshot soon after arrival.

2 Supermarine Southampton III K2964 was one of the two final Southamptons built and served with No. 204 between 1933 and 1935, no identifying markings apart from the serial number being carried.

3 One of the original Saro London Is, K5263 entered service with No. 204 Squadron in October 1936 and was passed on to No. 201 Squadron in August 1937.

© Hylton Lacy Publishers Limited

1. Avro Shackleton MR.2C WR966 carries No. 204's badge on the fins. Later this aircraft became a trainer with the Maritime Operational Training Unit.

2. Douglas Dakota C.4 KN654 was delivered to the Middle East in June 1947 where it was used by No. 204 Squadron. Returned to England at the end of the year, it was broken up for spares in February 1950. Post-war Dakotas in the Middle East normally had their camouflage paint removed and flew in natural finish in an attempt to reduce heat absorption.

3. Vickers Valetta C.1 VW165 was taken on charge by No. 204 Squadron on 15 September 1949 and remained with it until transferred to No. 84 Squadron on 20 February 1953. After a crash in Aden on 12 June 1957, VW165 was broken up for spares.

© Hylton Lacy Publishers Limited

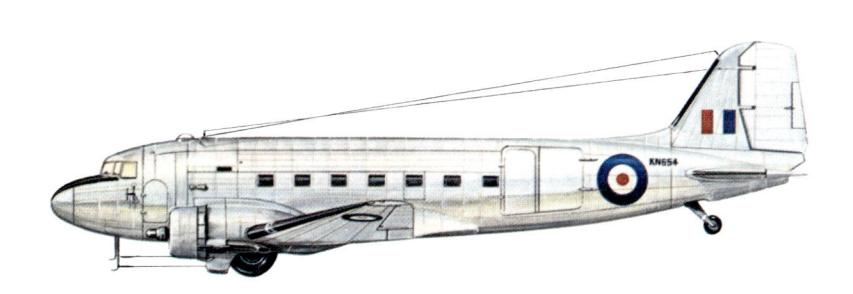

SQUADRON BASES

Bray Dunes, France	1 April 1918
Teteghem, France	13 April 1918
Cappelle, France	30 April 1918
Teteghem, France	9 May 1918
Heule, Belgium	24 October 1918
Waddington, Lincolnshire	7 February 1919 to 31 December 1919
Mount Batten, Devonshire	1 February 1929
Aboukir, Egypt	27 September 1935
Alexandria, Egypt	22 October 1935
Mount Batten, Devonshire	5 August 1936
Australian cruise	2 December 1937 to 29 May 1938
Sullom Voe, Zetland, Shetland Islands	2 April 1940
Reykjavik, Iceland	5 April 1941
Gibraltar	15 July 1941
Bathurst/Half Die, Gambia	28 August 1941
Jui, Sierra Leone	28 January 1944
Bathurst, Gambia	1 April 1944
Jui, Sierra Leone	8 April 1944 to 30 June 1945
Kabrit, Egypt	1 August 1947
Fayid, Egypt	22 February 1951 to 20 February 1953
Ballykelly, Co Londonderry, Ulster	1 January 1954
Honington, Suffolk	1 April 1971 to 28 April 1972

SQUADRON EQUIPMENT
Period of Use and Typical Serial and Code Letters

Sopwith Camel	April 1918 to February 1919	C71
Supermarine Southampton	February 1929 October 1935	S1037
Supermarine Scapa	August 1935 to February 1937	K4198
Saro London I, II	October 1936 to August 1939	K5911
Short Sunderland I	June 1939 to September 1943	L5800 (KG-J)
Short Sunderland II	June 1941 to March 1943	W3978
Short Sunderland III	October 1942 to June 1945	DD833 (KG-M)
Short Sunderland V	April 1945 to June 1945	ML872 (KN-B)
Douglas Dakota C.4	August 1947 to July 1949	KN654
Vickers Valetta C.1	May 1949 to February 1953	VW165 (D)
Hawker Siddeley (Avro) Shackleton M.R.1A	April 1958 to February 1960	WB860 (L)
Hawker Siddeley (Avro) Shackleton MR.2, MR.2C	January 1954 to date January 1954 to April 1972	WR966 (O) WR966 (O)

COMMANDING OFFICERS

Sqn Cdr B. L. Huskisson	1 April 1918
Major E. W. Norton	27 July 1918
Major L. S. Breadner	10 November 1918
Major E. W. Norton	21 November 1918
Major P. Huskisson	10 December 1918
Major R. S. Lucy	10 January 1919
S/Ldr F. H. Laurence MC	1 February 1929
S/Ldr K. B. Lloyd AFC	9 December 1930
S/Ldr A. W. Fletcher DFC, AFC, OBE	1 January 1934
S/Ldr V. P. Feather	1 October 1936
W/Cdr K. B. Lloyd AFC	19 October 1937
W/Cdr E. S. C. Davies	18 March 1940
W/Cdr K. F. T. Pickles	14 August 1940
W/Cdr D. I. Coote	22 May 1941
W/Cdr P. R. Hatfield	28 February 1943
W/Cdr C. E. V. Evison	24 March 1943
W/Cdr H. J. L. Hawkins	19 September 1943
W/Cdr A. Frame	17 August 1944
W/Cdr D. Michell	12 January 1945
S/Ldr H. S. Hartley	1 August 1947
S/Ldr R. A. Pegler	15 January 1948
S/Ldr L. W. Davies	22 May 1950
S/Ldr H. H. Jenkins	1 October 1952
S/Ldr G. Young	1 January 1954
W/Cdr W. Beringer	25 July 1955
W/Cdr A. D. Dart DSO, DFC	3 June 1957
W/Cdr J. C. W. Weller DFC	23 July 1958
W/Cdr R. D. Roe AFC	1 June 1960
W/Cdr C. K. N. Lloyd AFC	14 June 1962
W/Cdr J. J. Duncombe AFC	1 May 1964
W/Cdr P. Kent MBE	7 March 1966
W/Cdr O. G. Williams	17 June 1968
W/Cdr E. P. Wildy	14 April 1969
S/Ldr D. E. Leppard	1 April 1971

1 Shackleton 'O/204' carried the squadron badge on the fin. The wing 'stripes' result from engine exhaust soot *Photo: MoD. PRB25974)*

2 For major maintenance, Sunderlands were slipped on a beaching chassis. 'F/204' (T9072) is shown ashore at Bathurst undergoing maintenance. Leading-edge platforms are unshipped for work on the engines and the aircraft is picketed down to prevent damage in high winds. *(Photo: IWM. MH6670)*

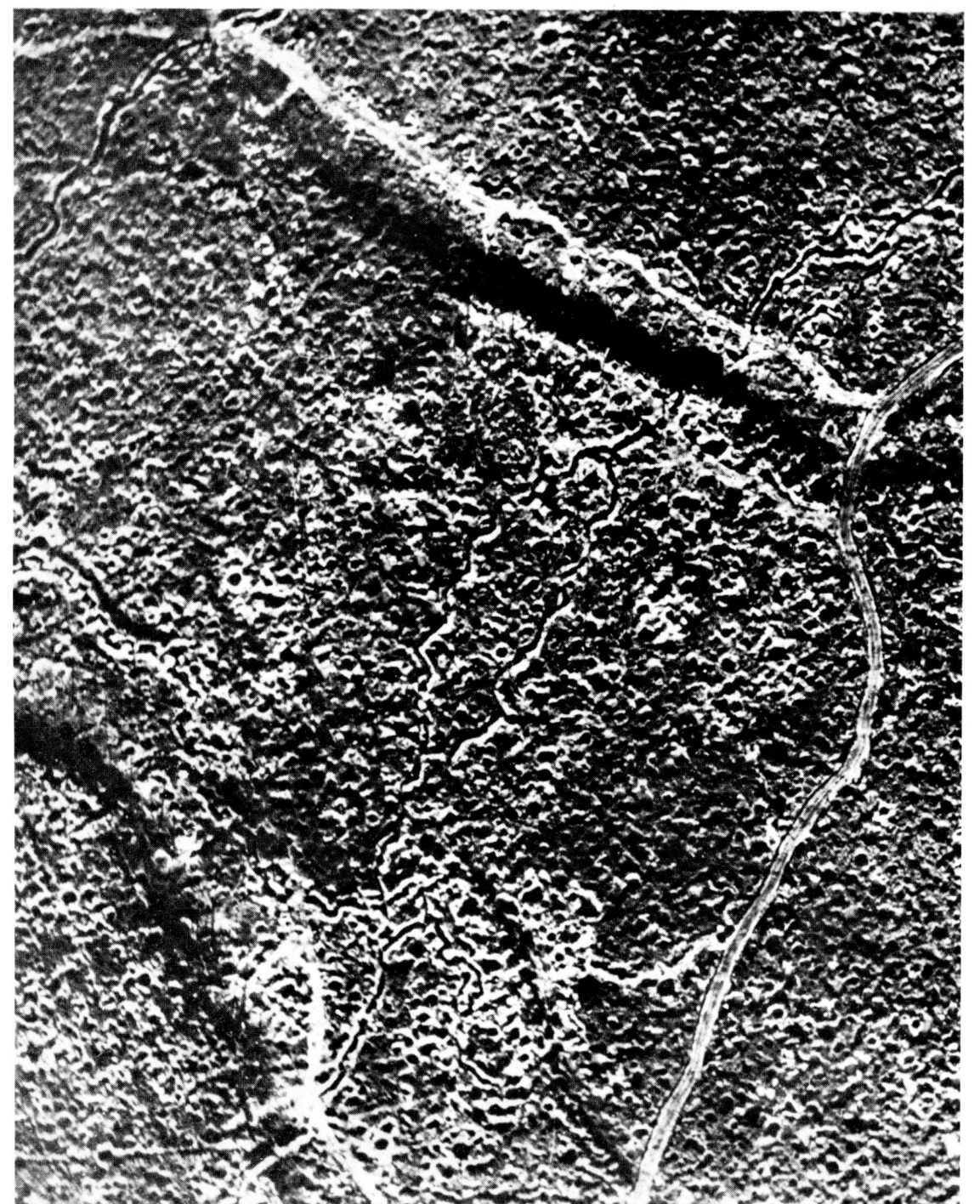

206 Squadron

Tactical Reconnaissance, September 1918: A section of trench lines near Ypres where interlocking shellholes have altered the landscape to a wasteland. Only the road has been kept clear for supply vehicles
(*Photo: Taken by No. 206 Squadron*)

The Photographers

Ever since the earliest days of photography, men have been taking pictures from the air. Primitive cameras were embarked in balloons with very variable results but it was not until World War One that photographs became an essential part of the support given to the army by aircraft.

In the opening months of the war, pilots and observers sketched the positions and dispositions of the enemy with varying degrees of skill. Several enthusiastic photographers were convinced that the camera could do the job better and experimental aircraft cameras began to be developed. By 1918, they had overcome such problems as aircraft vibration, freezing shutters, iced-up lenses and the innumerable snags that happen to cameras at the most awkward times. Photographing the front line was a routine task for the Royal Flying Corps squadrons.

Clustered around Dunkerque were the land-based squadrons of the Royal Naval Air Service. Though primarily intended to support naval operations off the Belgian coast, they were also giving support to the RFC which was of great value and many of the RNAS fighter squadrons made their reputations detached from naval affairs. There were also a few day bomber units attacking targets in Belgium and a further squadron began to form at Dover on 1 November 1917.

Manned by personnel from No. 11 Squadron, RNAS. and the Walmer Defence Flight, the new unit was designated No. 6 Squadron, RNAS. Its aerodrome was a small field outside Dover Castle which bordered on a much larger area occupied by the Royal Flying Corps. Within a few months, both services would be merged into the Royal Air Force but until then they kept to their own sides of the dividing hedge. In fact, the two airfields at Guston Road and St Margarets remained separate establishments until they were closed at the end of the war.

Under the command of Squadron-Commander (Acting) C. T. MacLaren, the squadron slowly acquired pilots and observers and began flying its Airco (de Havilland) D.H.9 bombers over the countryside of Kent. Built as a replacement for the earlier D.H.4, the DH9's greatest problem was its unreliable Siddeley Puma engine and the type was never popular with the day-bomber squadrons which had to set forth across the enemy lines in aircraft powered by an engine with a 10% failure rate per sortie.

In appearance, the D.H.9 appeared to be a useful addition to the bomber squadrons. Armed with two machine-guns, one fixed and forward-firing and one on the observer's gun-ring, it could carry a load of four 112-lb. bombs. Manoeuvrability was good for a day-bomber and, in contrast to the D.H.4, the pilot and observer sat back to back where they could communicate with each other instead of being separated by a large fuel tank.

No. 6 (Naval), as it was known in France to avoid confusion with No. 6 Squadron, Royal Flying Corps, left Dover on 14 January 1918 for the airfield at Petite Synthe on the outskirts of Dunkerque. There was still much work to be done to bring it up to strength. Not until 9 March did the squadron set off on its first bombing raid, the target being St Pierre Cappelle, near Dixmude. On the way back to base, a large enemy bomber was seen near the German lines where it had apparently made a forced landing. After rearming, the squadron returned to the position and damaged the enemy aircraft which was probably the Staaken R.VI *R27* which had just managed to reach Belgium after its fuel system had frozen and caused total engine failure just off the coast; it was finally destroyed by artillery fire. On the return flight, a Pfalz fighter which tried to attack the D.H.9 formation was shot down.

Other raids followed, notably on Bruges docks where several ships were hit; but, after only three weeks of operations with the RNAS, the squadron became part of a new service when the Royal Air Force was formed on 1 April 1918. From No. 5 Wing, RNAS, the squadron moved to the 11th Army Wing of the 2nd Brigade, RFC. Since all the naval squadron numbers duplicated those of the RFC, No. 6 (Naval) became No. 206 Squadron.

The squadron was soon on the move. The German offensive of March 1918 was still in full swing and

1 A standard D.H.9, the type No. 206 flew during World War One fitted with the unreliable Puma engine (*Photo: IWM. Q56858*)

2 The D.H.9's 'office' placed pilot and gunner close together to aid co-ordination between both members of the crew (*Photo: IWM. Q67315*)

3 and 4 Revealed in these photographs of 'A' and 'B' Flights at the end of World War One is the variety of uniforms found especially in former RNAS squadrons (*Photo: No. 206 Squadron collection*)

No. 206 moved back to undertake a new task. The Army Commander required up-to-date information on the enemy's positions and bases and for the rest of the war No. 206 played a major part in providing photographic coverage, taking almost 12,000 photos of a very high standard of the enemy lines facing the Ypres salient. The whole area was covered at least once a week to detect changes, and rear bases covered as required by lone, unescorted, D.H.9s.

Bombing raids continued to be flown in addition to the squadron's reconnaissance committments. These often met determined resistance from enemy fighters as when a flight of four D.H.9s were intercepted by 20 Pfalz fighters between Ypres and Courtrai. Captain H. Warren with his observer Lieutenant L. A. Christian (in B7596) was leading the formation and had the advantage of twin Lewis guns in the rear cockpit which accounted for two of the enemy seen to crash near Menin. Lieutenant G. A. Pitt and his observer Corporal G. Betteridge (in D1689) saw their adversary spin down smoking near Courtrai. Then two more enemy aircraft went down out of control under the fire of the other two D.H.9s flown by Lieutenant Burn with Captain Carrothers as observer (D5590) and Lieutenants Percival and Paget (D1015). Flexible twin Lewis guns were fitted on all squadron aircraft and proved their worth on many occasions.

The final Allied offensive that ended the war meant intensive work by all the RAF squadrons but few equalled No. 206's record of dropping 38 × 112-lb. and 299 × 25-lb. bombs in a single day's operations. However, these raids were not unopposed. On 30 October, for example, seven D.H.9s bombing Sotteghem were engaged by 20 Fokker D.VIIs. Close formation enabled the fire of all 14 Lewis guns to hold off the enemy and four Fokkers were destroyed without loss. By the time the enemy retired, two of the flight had been reduced to firing Very pistols—whose signal flares had been designed for more prosaic purposes.

With the Armistice on 11 November 1918, the squadron moved to Nivelles in Belgium for a short time before joining the Army of Occupation in Germany. In the course of 156 bombing raids, 116 tons of bombs had been dropped.

A total of 478 photographic missions had been flown and 26 enemy aircraft shot down; with 23 more being claimed as going down out of control. Twenty-six members of the air crews had been killed and 22 were missing of which 10 were taken prisoner. Thirteen more had been wounded.

After being based at Bickendorf on the outskirts of Cologne, No. 206 was posted to Egypt at a time when most squadrons were being transferred to England for disbandment. Arriving at Alexandria in June 1919, the squadron took up its base at Helwan and remained there until 1 February 1920 when it was renumbered 47 Squadron.

Coastal Patrol

As the expansion of the RAF in the mid-1930s began to produce a flow of aircraft, new squadrons began to form; among them was No. 206 Squadron. On 15 June, 1936, 'C' Flight of No. 48 Squadron at Manston, Kent, was detached to form a nucleus of No. 206. It brought with it six Avro Ansons as a first contribution to the planned establishment of 18, another six being kept as reserve aircraft. To man the squadron, 22 officers, 45 ncos and 113 airmen were authorized, 18 of the ncos being sergeant-pilots.

No. 48 Squadron was at that time engaged in training duties at Manston and was constantly expanding. It was destined to become the largest squadron ever to serve with the RAF because it also provided aircraft for the School of Air Navigation based at Manston. The sky over the Isle of Thanet was already sufficiently crowded for No. 206 to be posted to Bircham Newton in Norfolk and on 30 July Wing-Commander F. J. Vincent, DFC led a formation of seven Ansons northwards. On 4 August 1936, the squadron was formally embodied in Coastal Command as a general-reconnaissance unit.

During World War One, a number of squadrons of

1

landplanes had been formed to fly coastal anti-submarine patrols, supplementing the efforts of coastal seaplane stations. Most of their aircraft were D.H.6 trainers, augmented by a small number of D.H.9s withdrawn from day-bomber units. With the Armistice, the requirement for land-based patrol aircraft lapsed, this role being given to the handful of flying-boat units which remained in being. In 1934, the Air Ministry issued a requirement for a general-reconnaissance aircraft to operate from land bases; and, as a result of trials between a modification of the Avro Type 652 six-seat transport and a militarized de Havilland D.H.89 Dragon Rapide, the former was selected to fill this gap in the RAF's armoury.

An initial order of 174 aircraft was awarded to A. V. Roe & Co Ltd of Manchester, the military Type 652A being bestowed the name of Anson. Powered by two 350 h.p. Armstrong-Siddeley Cheetah radial engines, the Anson was a low-wing monoplane of mixed construction, the fuselage being metal and the wings and tail unit wood; fabric covering was adopted for the fuselage—or that part of it which was not already occupied by large windows. There was accommodation for a crew of three, pilot, observer/navigator and radio-operator/air gunner. Armament consisted of a 360-lb. bomb load in a bay in the centre section, a fixed 0·303in machine gun in the nose and a second mounted in the Armstrong Whitworth dorsal turret. The 'bird cage' turret was manually-operated and, when radio-operators were obviously going to be fully employed, an extra crew member was carried to act as gunner.

When Ansons began to reach No. 48 Squadron in March 1936, they attracted much attention from the Press, mainly because of the retractable undercarriage. As the first modern monoplane to enter service with the RAF, it was the subject of many inspections, articles and photographs and, by the standards of the mid-1930s, it was a very effective aircraft. Unfortunately, by the time war broke out, rapid development had rendered the type obsolescent—in common with many other advanced military designs of that period.

No. 206's Ansons were fitted with 295 h.p. Cheetah VIs as were many of the initial production Ansons pending the development of the more powerful mark of engine. On 14 September 1936, it began conversion courses for pilots emerging from the Flying Training Schools and destined for general-reconnaissance, flying-boat and bomber (Blenheim) squadrons. These courses continued until 6 June 1937 by which time a total of 278 pilots had successfully converted in 2,700 hours of flying and the cost of only one (repairable) accident.

Having put this mundane but essential task behind it, the squadron began to take part in exercises. Locating and shadowing ships at sea, practice bombing on the radio-controlled target ship *Centurion*, (a survivor of the battleships built just before the outbreak of World War One) and gunnery training occupied No. 206 until the Munich crisis.

With Germany threatening to invade Czechoslovakia, the War Defence Scheme was put into operation, all personnel recalled to duty and the silver-finished Ansons disappeared rapidly from the tarmac as camouflage was applied and code letters replaced the squadron number '206' carried on the fuselage of each Anson. Affiliation with the Spitfires of No. 19 (Fighter) Squadron, from Duxford in Cambridgeshire, provided practice for the squadron's gunners but left no illusions about the Anson's lack of speed and protection when faced with modern interceptors. Fortunately, it was believed, Ansons would operate out of range of single-seat fighters; a comforting but highly-inaccurate forecast.

The safety record of the Anson squadrons was high but No. 206 lost an aircraft (K8836) and crew on 1 November 1938 when it failed to return from a night navigation flight over the North Sea. No trace was ever found of the aircraft.

On 1 September 1939, General Mobilization was ordered as Germany invaded Poland. The same evening, four Ansons were despatched on a parallel track sweep over the North Sea for a distance of 110 miles from the Norfolk coast to locate shipping and

1 Strategic Reconnaissance, September 1918: Factories near Lille with supply dumps served by spur lines from the railway
(*Photo: Taken by No. 206 Squadron*)
2 Anson 'T/206' (K8754) in camouflage soon after the outbreak of World War Two
(*Photo: No. 206 Squadron collection*)
3 Hudson 'V/206' lifts its tail on take-off
(*Photo: No. 206 Squadron collection*)

search for possible German U-boats taking up position.

War Establishment was fixed at 16 Ansons with six in reserve and 24 aircraft were on hand. For an establishment of 26 officers and 247 ncos and airmen, the squadron was manned by 32 officers and 206 men. Under wartime conditions, aircraft could be expected to be lost or damaged in accidents at a far greater rate than hitherto and the war was only two days old when 'B/206' (K6183) was posted missing with its crew. On the same day, 'E/206' carried out a low-level bombing attack on a U-boat, dropping two 100-lb anti-submarine bombs without effect. Four days later, this aircraft was returning from a parallel-track sweep over the North Sea by a formation of 12 Ansons when it missed Bircham Newton in the black-out and became lost. Eventually it ditched in the English Channel and the crew were fortunate to be rescued.

Patrols were flown over coastal convoys and a detachment of six Ansons was sent to Carew Cheriton near Pembroke to co-operate with No 217 Squadron in flying patrols over the Bristol Channel and Irish Sea. One of these attacked a U-boat off Lundy Island on 20 September, again without effect. The 100-lb anti-submarine bomb in use had a very low lethal radius and under normal conditions had to explode within 10 feet of the U-boat to cause damage of any importance. Less than 1% of attacks by aircraft during the first year of war caused damage and it was some time before effective depth charges were available for RAF Coastal Command aircraft. The fate of HMS *Snapper* gave some indication of what the U-boats were suffering. The British submarine was subjected to an attack by a Coastal Command aircraft in December 1939 and received a direct hit from a 100-lb. bomb at the base of the conning tower—it was successful in breaking four light bulbs inside. Confidence in the RAF's primary anti-submarine weapon waned and the development of airborne depth charges accelerated.

With only a few GR squadrons available to cover the coastal waters of the British Isles, the available force had to be dispersed to suitable bases. Another detachment of No. 206 was sent to Hooton Park on the south shore of the Mersey. Squadron-Leader Hughes took over the station from No 610 Squadron, a local Auxiliary Air Force unit, and found little accommodation available for his seven Ansons and crews, the hangar and most of the buildings having been let to a civilian company. Nevertheless, patrols over the Irish Sea began next day.

During November, Bircham-based Ansons began to sight enemy aircraft. On the 7th, Pilot-Officer Henderson (in K6190) sighted and attacked a Dornier Do 18 flying-boat which retired after being hit in the wing. Flying-Officer Dias (in K6176) was at the receiving end of fire from a Heinkel He 115 seaplane and evaded it by diving into cloud. Next day, Pilot-Officers Greenhill and Featherstone, with Aircraftmen Britton and Gill, (in K6195), were on patrol 140 miles out over the North Sea when an He 115 seaplane appeared and opened fire. After firing 170 rounds back, the Anson crew saw the German two-motor seaplane fall into the sea. Soon afterwards, Pilot-Officer Harper's crew (in K6190) sighted a Do 18 on the water and bombed it, following this by several bursts from the front gun. It appeared to be damaged when a second Do 18 arrived and shots were exchanged until the Anson ran out of ammunition and returned to base.

At Hooton Park, reinforcements had arrived. Nine de Havilland Tiger Moth trainers were delivered for use by No. 3 Coastal Patrol Flight which was forming under the wing of No. 206 Squadron. The wheel had turned full cycle and once more unarmed trainers were to be used on 'scarecrow' patrols which acted on the principle that no U-boat commander was likely to remain surfaced when sighted by an aircraft. Once submerged, a submarine's visibility and mobility were restricted while naval craft could be directed to the position to hunt the U-boat. The flight became operational on 30 November under Pilot-Officer Stocken and No. 206 Squadron's detachment ceased to operate from Hooton Park. Four Ansons were taken over by No. 502 Squadron and the remainder of the detachment moved to Silloth in Cumberland to join the

RAF Coastal Command Pool and undertake conversion training on Lockheed Hudsons.

Out on patrol over the North Sea, an Anson (K6184) was 150 miles from base when it dive-bombed a surfaced submarine and reported a direct hit at the base of its conning tower. The crew of HMS *Snapper* confirmed the accuracy of this report with the results already mentioned. There was a marked increase in signals between Coastal Command and the Admiralty for a period.

Another Anson (K6189) failed to return from patrol on 6 December and was posted missing. For the rest of the winter, a particularly severe one, the squadron's aircraft patrolled over the North Sea searching for U-boats and escorting convoys despite snow, fog and gales. On 26 March 1940, the first signs of a change came with the arrival of two Hudsons of 'A' Flight from Silloth, followed by a trickle of Hudsons over the next few weeks. 'C' Flight became the only operational section of the squadron while armament training was carried out with the new aircraft but on 12 April two Hudsons (N7312 and N7343) were despatched on their first patrols.

The Hudson had arrived in RAF service as a result of a visit by an Air Ministry team to the USA in 1938. With expansion schemes becoming more and more ambitious, there was a gap appearing between the total aircraft production of the UK and the number of aircraft required to equip the squadrons and training units. Additional production capacity had to be found and Commonwealth countries began to build up their small aircraft industries. Foreign suppliers were considered but there were few sources which were neither too small to deliver in quantity nor potentially hostile. France had recently nationalized its aircraft industry and was currently producing more paper than metalware; the Netherlands' factories were small and committed to supplying France. Czechoslovakia would have been a logistic nightmare and Poland required every modern aircraft it could produce for its own air force. The Soviet Union could scarcely be considered and Germany was unlikely to oblige. Which left Italy and the United States of America.

Despite the apparent closeness of the German and Italian regimes, negotiations were put in hand for the purchase of Caproni general-purpose aircraft of similar design to the Anson. Fortunately, these did not reach fruition as the type selected proved to be deficient in several respects, notably that of safety.

In the USA there was already a number of major companies engaged in the manufacture of aircraft whose production capacity. A team was despatched to investigate the sources of supply and were directed to select two types, one a general-reconnaissance aircraft and the other a pilot trainer. After inspecting such types as the Douglas B-18 (a bomber developed from the DC-3), Martin 166, Boeing B-17 ('Flying Fortress') and a proposed amphibious version of the Consolidated PBY, the group arrived at the Lockheed plant. There they found a mock-up of a military version of the Lockheed Model 14 transport. Of all-metal, stressed-skin construction, this example of the Lockheed transport series impressed the British visitors more than anything they had seen previously. It had adequate speed and range to act as an Anson replacement while the Boeing, Martin and Douglas submissions were unsuited to the role envisaged. Consolidated's amphibious Catalina was not considered practicable; nevertheless, it eventually appeared.

After inspecting Lockheed's mock-up, various modifications were suggested which permitted easier crew movement and repositioned the navigator in the nose. Rapid work incorporated these changes before the team's next visit and after further conferences the Lockheed design (and a trainer from North American —the NA-16 that became the famous Harvard) was recommended for purchase.

An initial order for 200 was placed soon afterwards and the first aircraft for the RAF came off the production line in December 1938. From February 1939 a flow of Hudsons, as the aircraft had been named, began to arrive in Britain where they were fitted with Boulton Paul dorsal turrets carrying two 0·303-in. Browning machine-guns. Two more guns were mounted in the nose and a fifth in a ventral mount. Two more could be fitted in beam positions making the Hudson far more defensible than the Anson. With a maximum speed of 246 m.p.h., the Hudson had a range of over 2000 miles, more than twice that of the Anson.

In 1940, No. 206's Hudsons were fitted with what was to become an essential piece of equipment for maritime aircraft. This was a radar set for locating ships designated ASV (Air-to-Surface Vessel) which were installed at St Athan during April. Patrols began to be flown along the German coast carrying three 250-lb. bombs in search of German shipping and on 1 May the squadron's first attack on enemy ships took place. Flight-Lieutenant Biddell and his crew (in N7351) dropped three bombs on two ships off the Friesian Islands. Meanwhile, Ansons continued anti-submarine patrols.

Frequent tours of the German coast could not expect to remain unmolested. On 2 May, Squadron-Leader Carr (flying N7300) was intercepted by a Messerschmitt Bf 109 E over the Elbe estuary but managed to reach cloud cover with only a single bullet hole in his Hudson. Next morning, Pilot-Officer Kean (in N7319) was passing Norderney when three Bf 109s attacked from astern. The Hudson dived to sea-level amid a hail of cannon shells and bullets which killed the gunner, Leading Aircraftman Townsend. But return fire hit the leading Messerschmitt and it was seen going down steeply towards the sea. Flying a few feet above the waves, the remaining enemy fighters were shaken off and N7319 brought home to a crash-landing at Bircham Newton, the result of a damaged hydraulic system.

German troops rolled over the frontiers of Holland and Belgiums on 10 May and No. 206 despatched five Hudsons on reconnaissance flights along the Dutch and German coasts. Two days later, Pilot-Officer Gray and crew (in N7353) failed to return from a similar mission. Two more Hudsons (N7329 and N7400) were lost after they joined others from Thornaby in bombing Hamburg on 18 May. On the 22, another Hudson (N7402) did not return from patrol. From 28 May, formations of three Hudsons flew standing patrols over shipping engaged in evacuating Allied troops from Dunkerque.

One patrol of Hudsons found a formation of six Bf 109s attacking Royal Navy Blackburn Skua dive-bombers on the evening of 3 June. Flight-Lieutenant Biddell in N7351 led his two wingmen (Pilot Officer Kean in N7333 and Flying Officer Marvin in P5133) into the melée and the Messerschmitts broke-off action after losing two of their number. Biddell was awarded the Distinguished Flying Cross a few weeks later. On 19 June, he was the pilot of the Hudson which evacuated General Sikorski and his staff from Bordeaux to England where a Polish army was formed.

During June, the last Ansons were posted away having flown their final patrols on 25 May. Hudsons were used for anti-shipping patrols and bombing attacks on coastal targets. On 21 June, Pilot-Officer Featherstone (in P5153) located a convoy of five supply ships and three escorts off Texel and this sighting resulted in nine Hudsons setting off to attack it in the early afternoon. Failing to find the convoy, the formation bombed shipping at Den Helder. The nearby airfield at de Kooy was visited by six Hudsons of 'C' Flight on 24 June when many hits were claimed on hangars and buildings.

Two more Hudsons were lost when N7368 and P5162 failed to return from a search for a Handley Page Hampden of No. 44 Squadron which had 'ditched' off Texel.

Occasional encounters with enemy aircraft were indecisive. An attack on a floatplane He 115 C by N2396 on 27 August resulted in the enemy aircraft diving to sea level. Another Hudson sighted the seaplane on the water but as it turned to machine-gun the Heinkel, it lost sight of it in the gloom. The He 115 regained its base with the crew of three wounded. Another which crashed on landing in Amsterdam

1 Hudson 'A/206' (T9444) demonstrates the effectiveness of its camouflage pattern over a checkerboard of fields in Lincolnshire (*Photo: IWM. CH5554*)

2 'C/206' shows off the Hudson's bulbous Boulton-Paul turret fitted after delivery of the aircraft from the USA (*Photo: IWM. CH282*)

harbour on 2 October may have been one of a pair of Heinkel floatplanes attacked by a Hudson while minelaying.

Some of the earliest 'Intruder' operations were flown by No. 206 in November. Under the code-name *Race*, the first of these was flown by Pilot-Officer Ward (in N7318) on 14 November. Locating an airfield near Abbeville, three attacks were carried out on each of which a 250-lb. bomb was dropped. While climbing away after the third run, a Messerschmitt Bf 110 was seen taking-off and as it came in astern, Sergeant Garrity fired 600 rounds from his turret. The enemy fighter spun into a wood.

In March 1941, the round of shipping reconnaissance and occasional bombing raids was interrupted by the despatch of a detachment to Aldergrove in Northern Ireland. The U-boat menace was increasing and more patrol aircraft were required. After six weeks, the detachment returned; but, on 6 May, another flight of Hudsons was sent to St Eval in Cornwall where it was joined by the remainder of the squadron at the end of the month. One of the first tasks given to No. 206 was the location of the German battleship *Bismarck* and the heavy cruiser *Prinz Eugen* approaching Brest. The heavy cruiser evaded the patrols and reached harbour but its giant consort was already at the bottom of the Atlantic.

German U-boats were based in bomb-proof submarine pens at Brest, Lorient, St Nazaire and La Pallice on the west coast of France. Coastal Command patrols covered the approaches to these ports and No. 206 flew many sweeps over the Bay of Biscay in the hope of intercepting U-boats on their passage to and from their operational areas. During one patrol 'B/206' (AE613) landed in the sea after its starboard engine caught fire and sank in under two minutes, leaving Pilot-Officer Kennan and his crew in a rubber dinghy. Next day, 'M/206' sighted the dinghy at mid-day and a few hours later Sunderland 'L' (T9047) of No. 10 Squadron, Royal Australian Air Force, circled overhead. As it touched down, a heavy sea damaged its port engines and the big flying-boat wallowed helplessly in the swell. More help was on the way, however, and shortly after midnight on 10 July two destroyers

1 On the right, and in front of a No. 206 Squadron Hudson, is Acting Wing Commander Lionel Cohen, DSO, MC, who won the DFC in January 1944 when over 60 years of age. While Air Liaison Officer to the Royal Navy, he flew many sorties as an air gunner or observer (*Photo: IWM. CH12099*)

2 Hudson 'B/204' displays to good advantage the portly lines of the Lockheed design (*Photo: IWM. CH6573*)

3 'Q/206' undergoes maintenance in 1942 (*Photo: IWM. CH5799*)

4 Fortress II 'J/206' (FL459) taxies out on the windswept airfield at Benbecula. Note the underwing ASV aerials (*Photo: IWM. CH11131*)

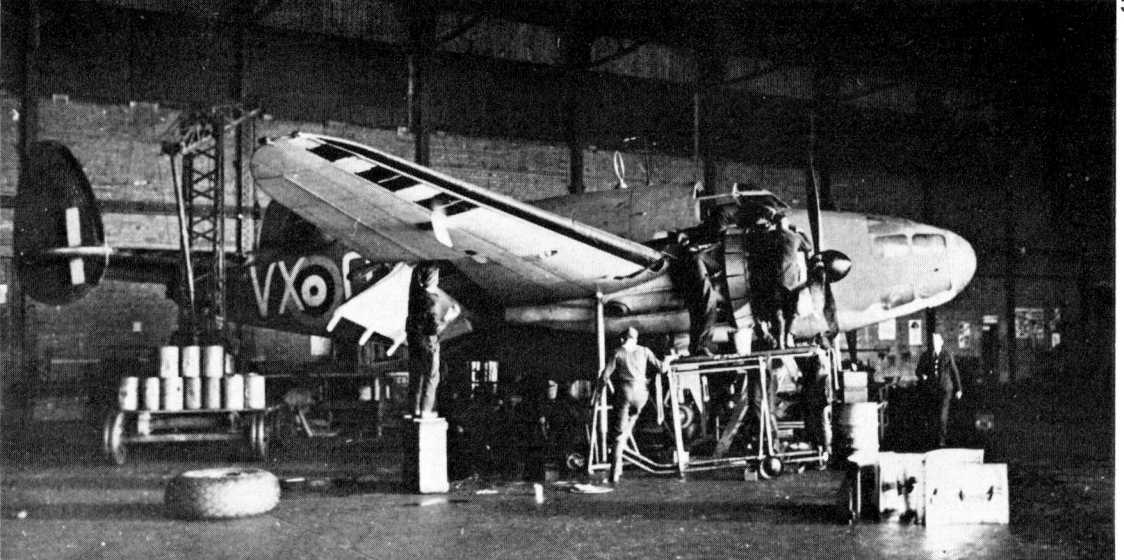

arrived and picked up both crews, sinking the Sunderland by gunfire.

Occasionally attacks were made on coastal shipping with varying results. One of the most effective was when Sergeant Whitfield in 'E/206' (AM602) placed three 250-lb. bombs aboard a large minesweeper after a sea-level approach in the face of heavy *flak* on 11 July.

The scene of operations changed in August when No. 206 moved to Aldergrove to patrol the approaches to the British Isles. Long patrols in all types of weather became the daily round for the squadron and many aircraft operated away from their home base while covering convoys over the stormy waters of the North Atlantic. A shuttle patrol between Wick and Reykjavik in Iceland was set up and another flew in a wide sweep over the Atlantic between Aldergrove and Stornoway in the Outer Hebrides.

One of the last tasks given to No. 206's Hudsons before they were replaced was a directive to supply 12 aircraft to take part in one of the 'Thousand Bomber Raids' in June 1942. Taking-off from Donna Nook in Lincolnshire for Bremen, they carried out their missions with the loss of two Hudsons. Both 'M/206' and 'S/206' were missing, the latter with the squadron's Commanding Officer, Wing-Commander H. D. Cooke, aboard. Both aircraft apparently landed in the North Sea on their return flights.

On 20 July, conversion training from some of No. 206's crews began on Boeing Fortresses of No. 220 Squadron. Longer-range aircraft were required to fly further and longer over the Atlantic and the faithful Hudsons were about to be replaced by four-engined Fortresses. Ironically, this was one of the types rejected by the British Purchasing Commission and the experiences of No 90 Squadron in operating improved versions as day-bombers had borne out the wisdom of the Commission. The surviving Fortress Is had been passed on to RAF Coastal Command but No. 206 was scheduled to receive the improved Fortress II.

The original 'Flying Fortress' had been heralded by unprecedented publicity as the height of American aeronautical engineering skill. In construction it was advanced but as a combat aircraft it was deficient in many respects. Its armament of five hand-held machine-guns belied its name, its bomb-load was small by RAF standards for an aircraft of its size and it was wasteful of trained aircrew. By comparison, the least-effective of the new generation of RAF bombers, the Short Stirling, was well-armed (three power-operated gun turrets carrying eight guns as against five hand-held guns), had a heavy bomb-load (up to 14,000 lb. compared to 2500 lb.) and had a crew of seven instead of 10. The Fortress was 50 m.p.h. faster and had a far higher service ceiling than the Stirling but neither could prevent it falling victim to enemy fighters.

The Fortress II, equivalent to the US Army Air Corps' (later Air Forces') B-17E, was a much-improved version and became the backbone of the USAAF Eighth Air Force when it began operating from Britain. Armament had increased to 13 guns, and initially there were two power-operated turrets, a third being fitted later below the nose. Armour had been installed and although the maximum speed had dropped considerably, it had an increased bomb-load and still retained its long range.

A handful of Coastal Command Consolidated Liberators had been attempting to fill the gap in mid-Atlantic between the maximum patrol radii of British-, Icelandic- and Canadian-based aircraft. To supplement these, more Fortresses and Liberators were ordered against stiff opposition from the US Army which regarded all bombers as their exclusive property. Bound to a policy of using Fortresses and Liberators for day bombing, any other use for them was sacrilege to the military staff, but despite their efforts an allocation of long-range aircraft for RAF Coastal Command and the US Navy was forced through.

Fortresses for the squadron began arriving on 29 July 1942 and the last Hudson patrol (by 'O/206': AM734) was flown on 27 August. At the end of June, No. 206 had moved to the bleak surroundings of Benbecula in the Outer Hebrides and here the squadron completed its conversion training. By the time 'A/206' flew its first patrol on 19 September, most of the work had been done and the squadron became fully operational on 1 October.

Normally, Fortresses carried 14×300-lb depth-charges but to increase the duration of patrols, seven of these were replaced by a 340-gallon tank in the bomb-bay; which gave nearly three hours extra cruising. Misfortune struck within a few days when a Fortress ('J/206': FL454) crashed into the sea immediately after take-off with the loss of five of its crew.

After years of fruitless patrolling in Hudsons, success with the new aircraft came quickly. On 27 October, four Fortresses were escorting convoy *SC 105* inbound from Sydney, Nova Scotia, when 'F/206' (FL457) sighted a U-boat on the surface. Running in over the enemy ship, seven depth charges were dropped a short distance ahead of the swirl left behind by the diving submarine. Pilot-Officer Cowey and his crew circled a patch of oil which marked the end of *U-627*; then resumed patrol.

Several more sightings and attacks were made before the end of the year but 'A/206' (FL453) and its crew failed to return to base after covering convoy *ONS 162* on 14 December. Three Fortresses swept the area without success. A month later, a pair of Fortresses were covering convoys *ONS 160* and *ONS 161* when 'C/206' (FK213) sighted a U-boat two miles to starboard but by the time the aircraft reached the spot it had dived. 'G/206' (FL452) manned by Pilot-Officer L. G. Clark and his crew surprised a U-boat, possibly the same one, and though three of the depth charges did not release, the remaining four straddled the target. The stern of the submarine rose to a steep angle and remained visible for a short time before the U-boat slid below the surface. The victim, *U-337* never returned to its base.

Two more U-boats were accounted for in March 1943 by one Fortress 'L/206'. Flown by Flight-Lieutenant W. Roxburgh's crew (on the first occasion) on 25 March, FK195 was five hours out from Benbecula when a U-boat was sighted five miles away. Six depth charges straddled the submarine perfectly. As its stern rose high in the air, the Fortress dropped its remaining depth charge as *U-489* disappeared to the bottom of the Atlantic.

Two days later, the same 'L/206' was back on patrol, this time with Flying-Officer A. C. I. Samuel and his crew. When the U-boat was sighted on the surface three miles away, it opened fire with light *flak*, fortunately inaccurately. Six charges landed close by and it heeled over to starboard. Fifteen seconds after the U-boat submerged, its bow reappeared and the final depth charge was dropped into the patch of foam. Before vanishing, men were seen scrambling out of the conning tower and thirty seconds after the final attack, *U-169* sank.

Yet another U-boat, *U-710* met its end on 24 April when Flying-Officer R. L. Cowey and crew (in 'D/206': FL451) caught it on the surface near convoy *ONS 5*. Next to go was *U-417* on 11 June, sighted from seven miles away by 'R/206' (FA704) flown by the Command-Officer, Wing-Commander R. B. Thomson, DSO.

This time, however, return fire was accurate. The Fortress was hit in the nose, cockpit, wings, bomb-bay and rear turret but the attack was pressed home and depth charges straddled *U-710* from 50 feet. The stricken U-boat slid below the surface leaving 20 to 30 survivors in the water. But 'R/206' was also near its end. With one engine stopped and another losing power, the Fortress was unable to maintain height and had to be 'ditched'. Before 'R/206' sank, the entire crew took to the dinghy—including the Station Armament Officer, who had picked an unfortunate mission on which to fly.

Later, three Fortresses set out on a search for the missing crew of 'R/206'. But, it was the next day before 'B/206' (FK208) sighted the dinghy—in the waste of water between the Faroes and Iceland—and dropped a pair of 'Bircham Barrels' containing survival equipment before making for Iceland low on fuel. On the 13th, three

1 Lockheed Hudson I (serial number T9303, obliterated under the 'X' when code letters were added) of No. 206 Squadron. Another 'V-VX' (P5162), in company with N7368, was despatched on 4 July 1940 to search for a Hampden of No. 44 Squadron down in the sea off Texel and both failed to return.

2 Airco (de Havilland) D.H.9 D7222 was in use with No. 206 Squadron during the final months of the First World War and unlike most other squadrons' aircraft carried no individual squadron or aircraft identification letter or mark.

3 Avro Anson I K6189 was delivered to No. 206 Squadron in August 1936 and was flown until 6 December 1939 when it failed to return from patrol and was posted missing with its crew.

© Hylton Lacy Publishers Limited

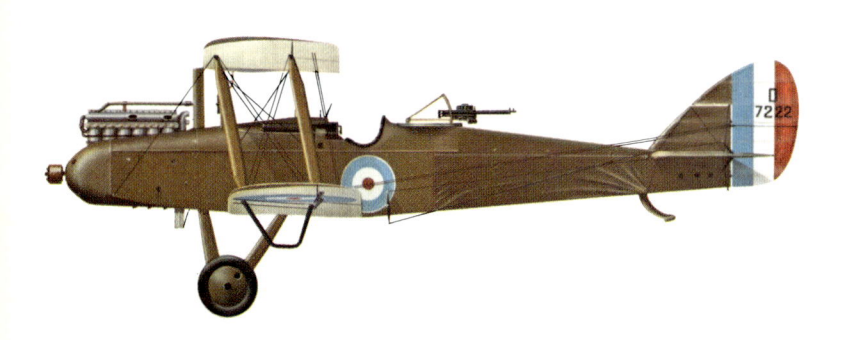

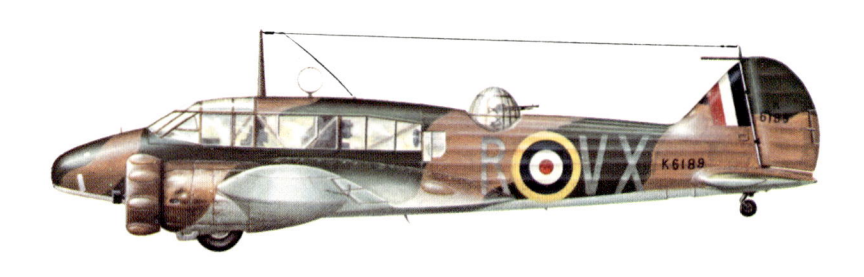

1 Boeing Fortress IIA FL459 was one of a group of Fortresses operated by Coastal Command from the Azores and flown by crews from Nos. 206 and 220 Squadrons. After serving with Nos. 519 and 251 Squadrons in 1945, FL459 was scrapped in December 1945.

2 Consolidated Liberator GR.VI EV885 was delivered to No. 206 Squadron in May 1944 but failed to return from patrol on 28 September 1944.

3 Avro Shackleton MR.1 WG526 was flown by No. 206 Squadron between July 1954 and November 1956. The squadron badge appears forward of the cockpit.

© Hylton Lacy Publishers Limited

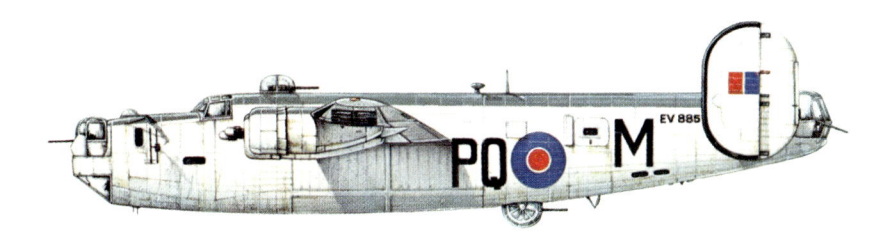

Fortresses failed to relocate the dinghy but another Fortress found it again early on the 14th and dropped a radio and more supplies. Help was gathering and when the aircraft left the area four hours later, three Hampdens, a Catalina and another Fortress were on hand. Shortly after mid-day, a Catalina ('L/190': FP102) from No. 190 Squadron's base at Sullom Voe in the Shetlands landed to pick up the castaways. It was an epic rescue, requiring the location of a tiny dinghy somewhere in thousands of square miles of ocean.

Long-range Liberators had begun to fill the patrol gap in the northern half of the Atlantic. Further south a new base was about to be acquired in the Azores under a long-forgotten treaty with Portugal which was resurrected as a convenient reason for British aircraft to operate from neutral territory. Nos. 206 and 220 Squadrons were designated for transfer; and, on 1 October 1943, ground crews sailed for the islands where they arrived on 8 October. The only available airfield at Lagens was barely adequate for big, four-motor aircraft. Nevertheless, on 18 October, Wing-Commander Thomson arrived in Fortress 'M/206'. Three days later, the first patrol was flown. By the end of the month, the rest of the squadron was operational from its new home.

Patrols radiated from the islands to join up with the patrol areas of other units based in Gibraltar, Morocco and the West Indies. The isolation of the Azores base was illustrated on 29 November when 'B/206' (FK208) was diverted to Gibraltar only to find the airfield closed by fog. With insufficient fuel to reach Port Lyautey in Morocco, the Fortress 'ditched' off Carnero Point with the loss of its crew of eight. Another Fortress, 'U/206' (FA705) failed to return from patrol and was believed to have fallen victim to *flak* from a U-boat.

Not before March 1944 was another U-boat sunk. Flight-Lieutenant A. D. Beaty and crew (in 'R/206': FA700) had sighted markers dropped by a Wellington of No. 172 Squadron after the latter had attacked a U-boat. As 'R/206' circled the spot, a surfaced submarine was sighted—one mile south of the markers —and opened fire. Then the submarine fired three red flares. The Fortress responded with a two-star cartridge. Fire ceased, presumably while the U-boat tried to sort out the differences between a Fortress and a Focke-Wulf FW 200 Condor. As the British aircraft made its obviously hostile run-in on the target, *flak* opened up again. But four depth charges landed close aboard *U-575* and she sank at a steep angle. Credit for the sinking was shared between No. 206 and other aircraft and ships which had already hunted the U-boat.

Six days later, on 19 March, movement orders reached the squadron. Patrols had ceased on the 16th in anticipation of a change of base; and, on 23 March, personnel began leaving the Azores by air for Prestwick in Ayrshire where they were ordered to Davidstow Moor in Cornwall. The remaining ground personnel joined them until the last nine members of No. 206 left on 14 April. The squadron's Fortresses had been flown back to Britain and their crews were sent to No. 1674 Heavy Conversion Unit at Aldergrove in Northern Ireland for conversion training on Liberators. By the end of April, nine Liberator VIs had been collected from Prestwick and Lichfield.

Although the Consolidated Liberator had not made its first flight until almost four years after the Fortress, it had a similar performance to the later marks. RAF Coastal Command had received its first Liberators in June 1941; and, if the type was more vulnerable to anti-aircraft fire and fighters than the Fortress, it was a better reconnaissance aircraft—being roomier and of longer range. The Mark VI version which No. 206 received was fitted with nose and ventral turrets in addition to the standard dorsal and tail turrets and also had an airborne searchlight fitted under the starboard wing. This enabled U-boats to be stalked on radar at night and illuminated for the final stages of the attack.

As the Liberators arrived, they were despatched to St Eval, further down the coast and the remainder of the squadron had completed their move by 12 April. Wing-Commander A. de V. Leach, DFC, took out 'B/206' (BZ962) on the squadron's first patrol over the Bay of Biscay on 23 May. To augment patrols along the Norwegian coast, a small detachment was sent to Tain in northern Scotland.

It was not long before results were seen. On 10 June, four days after the Allies had landed on the Normandy beaches, Flight-Lieutenant A. D. S. Dundas and crew (in 'K/206': EV943) were patrolling the western flank of the invasion area when a radar contact turned out to be a U-boat on the surface with a Mosquito overhead carrying out dummy attacks to keep it occupied. The Liberator ran in over the target and dropped five depth charges with no apparent results. The next six straddled the U-boat which slid down stern first, leaving three of the crew in the water. Nevertheless, no U-boat loss could be traced to this episode.

On 8 July, orders were received to move to Leuchars in Fifeshire. Allied armies had invested or captured the Biscay ports by September but even two months before the U-boats were abandoning their French bases. The Liberators left St Eval on 11 July and three days later six aircraft set out to patrol off Norway, one of which made an inconclusive attack on a U-boat which had just submerged. Next day, eight Liberators were despatched and Flying-Officer B. W. Thynne and crew (in 'E/206': EV947) were lost during an attack which sank *U-319* about 100 miles west of the Jutland coast. 'B/206' located a dinghy and one occupant near a very large oil patch and dropped a radio and supply pack but the survivor made no attempt to retrieve them. Later, Vickers-Armstrongs Warwicks dropped a lifeboat and 'Lindholme' gear without result. No further trace was found of the Liberator or any more of its crew.

Another Liberator ('D/206': EV873) was lost on 20 July when it failed to become airborne on take-off with the loss of eight of its crew. Then 'B/206' crashed on a flight between Tain and Leuchars on 14 September. On the other side of the ledger, one of the squadron's Liberators was on patrol between Norway and the Shetlands when Flying-Officer P. F. Carlisle and his crew (in 'S/206': BZ984) engaged a surfaced U-boat. Its *flak* was silenced by the nose gunner on the run-in and seven depth charges ended the career of *U-865*. Nine days later Carlisle and his crew were lost on patrol when 'M/206' (EV885) failed to return to base.

The Liberators' sweeps along the Norwegian coast had not gone unnoticed by the *Luftwaffe*. On 15 November, 'D/206' was attacked from dead astern by a trio of Messerschmitt Bf 110s off Bergen. The Liberator was flying at 150 feet and headed westwards when cannon shells hit one engine and wrecked the hydraulic system. During successive attacks, the port beam gunner was killed and the rear gunner was wounded. But, on the ninth pass, the mid-upper gunner hit one of the enemy fighters and it retired with an engine smoking accompanied by the other Messerschmitts. 'D/206' crash-landed at Sumburgh in the Shetlands without further injury to its crew.

Night sweeps off Norway began in February 1945. On 3 February, four Liberators were engaged. 'C/206' made a radar contact on an enemy coaster and dropped charges alongside; later still, straddling a U-boat with its remaining depth charges. 'J/206' also dropped on a surfaced U-boat and 'E/206' made a radar contact. Running in, the Liberator exposed its searchlight and illuminated a heavily-armed *Narvik*-class destroyer. As the nose turret replied to the *flak* that curved up from the warship, six depth charges dropped towards the enemy and the aircraft retired with many holes but no casualties.

In March 1945, re-equipment of the squadron with Liberator VIIIs equipped with Mark X radar began. Patrols ranged farther afield as the bases open to U-boats contracted. On 18 April, Flight-Lieutenant G. R. Haggas and crew (in 'J/206': KK250) attacked a surfaced U-boat off the Firth of Forth with the aid of sonobuoys while on 5 May, Flight-Lieutenant G. B. H. Thompson and his crew sank *U-534* while flying 'T/206' (KK250) over the Kattegat to the south of Goteborg. This was the tenth and last to be sunk by the squadron and a few days later it was engaged in escorting U-boats into British ports. In case the word had not reached some U-boats, patrols were continued until 3 June. A week later, No. 206 was given a new role.

---

1 Fortress (FK190) carries four ASV radar masts on top of the fuselage and additional antennae on the nose (*Photo: IWM. CH11130*)

2 Fortress FA702 was 'P/206' during service in the Azores (*Photo: No. 206 Squadron collection*)

3 Fortress 'H/206' takes off from dusty Lagens airfield in the Azores (*Photo: IWM. CA9*)

4 Catalina 'L/190' retrieves the crew of 'R/206' in the North Atlantic on 14 June 1943 (*Photo: IWM. C3606*)

5 Wing Commander Thomson and the crew of 'R/206' in their dinghy as the rescuing Catalina prepares to pick them up (*Photo: IWM .C3605*)

The Pantechnicons

On 10 June 1945, No. 206 Squadron found itself transferred to RAF Transport Command as part of No 301 Wing. In the following weeks, turrets were removed from its Liberators and the gaps faired over. Bomb-bays were sealed and search radar and other maritime equipment were stripped to leave a bare cavernous interior. Attachment points were fitted for seats or for cargo lashings and the squadron was reorganized. Crews were in future to consist of two pilots, one navigator, one radio operator and one flight engineer, a total of 48 aircrew being allotted to No. 206.

At the beginning of August, the squadron moved into its new home at Oakington in Cambridgeshire, where the Liberators began long-distance trooping flights. No. 206's terminal was at Karachi, now capital of West Pakistan, where the war in the Far East was coming to a close. When the squadron had been allocated to Transport Command, its task had been foreseen as being mainly the transport of troops and equipment to India. The dropping of atomic bombs on Japan in August 1945 had unexpectedly reversed the flow though a constant stream of replacements was flown out to replace men who had done more than their share of service in South-East Asia Command.

A considerable number of squadrons had been transferred from RAF Bomber and Coastal Commands to trooping duties and with the onset of demobilization, a manpower problem developed. Many were disbanded as the initial wave of trooping receded. As a result, on 25 April 1946, No. 206 ended its wartime career by being disbanded.

Logically, it should have been reformed as a maritime squadron but such units were few and those in being had already been renumbered with traditional 'maritime' numbers. No. 206 remained as a 'number-plate' until such time as a new squadron could be given its number. It was not long before the need for new squadrons arose.

On 17 November 1947, No. 206 reformed at the Transport Command base at Lyneham in Wiltshire. Its equipment was the Avro York—a bulky, high-wing monoplane powered by four Rolls-Royce Merlin XX engines which had been based on the Lancaster bomber. During the war, the development of transport aircraft by Britain had been brought to a halt by the need for combat aircraft. An agreement between Britain and the USA had resulted in all transport aircraft development being carried out in America and Dakotas, Liberators and Skymasters were delivered under Lend-Lease. This had resulted in the American aircraft industry having almost a clear field when airlines resumed civil flying in the years immediately after the end of the war. Time was required to develop national designs and it was to prove difficult to persuade airlines to change their equipment once they had standardized on the only types available when the air routes were re-opened.

The York was the exception to the rule. By using Lancaster wing and tail design, a new fuselage could be developed without diverting resources from the production of bombers. Permission was given to proceed with design work and the construction of prototypes on condition that it did not interfere with Lancaster production as it was foreseen that the end of the war was likely to end in the abrupt cessation of Lend-Lease supplies and many American-built aircraft would be grounded through lack of spares. With much of Britain's dollar resources spent in purchasing munitions from the USA, there would be a shortage of foreign currency to buy such items and some insurance against this probability was necessary.

The prototype York made its first flight on 5 July 1942 but only a handful were delivered by the end of the war due to lack of priority for non-combat aircraft. These were used mainly for VIP use but production built up when peace returned and 257 were built. Though outclassed by types designed primarily as airliners, the York carried a useful load and had a reasonable performance.

When the Russian blockade of Berlin began in June 1948, Transport Command faced a task which strained all its resources. Initially, Dakotas of No. 46 Group began an airlift from Wunstorf in the British Zone of Germany and runways were improved at Gatow in the British Zone of Berlin to supplement pre-war Tempelhof Airport in the American Zone. When the new runway became operational in July, the Yorks of No. 47 Group began flights into Berlin with supplies for the city, on the return flight bringing out manufactured goods and passengers. Seven squadrons, one of them No. 206, were engaged on the Berlin Airlift with Yorks, aided by Dakotas and Hastings, and by the end of the operation, the RAF transport squadrons had flown 65,857 sorties which carried 394,509 tons of material.

The Russian blockade collapsed in May 1948 though the organization which had defeated it was kept running for a period. No. 206 had seen little of its home base during the operation as its Yorks operated from Wunstorf and only returned to Lyneham for major overhauls. Now that they were back from Germany, the squadron's aircraft flew freight and passenger flights to RAF bases overseas until the squadron was again disbanded on 20 February 1950.

Return to the Sea

In September 1952, just over 2½ years later, No. 206 rejoined the ranks of the maritime squadrons. On 27th of the month, the squadron reformed at St Eval in Cornwall and began working up with Avro Shackletons. Development of this development of the Lincoln bomber had been protracted but in 1951-52, six squadrons were equipped. Its four Rolls-Royce Griffon engines gave the Maritime Reconnaisance Mark 1 Shackleton a useful range and 77 were built before production switched to the improved Shackleton MR.2.

Maritime patrol and air-sea rescue missions were the squadron's main tasks from the UK but many detachments were provided for overseas service. After

1 Liberator VIII 'L/206' (KK335) was the squadron's final wartime type
(*Photo: No. 206 Squadron collection*)

2 Shackleton MR.3 'C/206' (XF707) flying over the Cornish coastline (*Photo: Air-Britain archives*)

3 Liberator VI 'Y/206' (EV828) in the snow after conversion to transport duties. Turrets have been removed and fairings fitted, all maritime equipment being stripped to make way for seats or freight
(*Photo: IWM. HU3037*)

4 Shackleton 'F/206' taxies out at Kinloss. The two 20-mm. nose cannon have been removed
(*Photo: MoD. PRB18247*)

5 The impressively-bulbous nose of the Shackleton MR.1 housed search radar. The Commanding Officer's pennant is carried above and behind the squadron badge
(*Photo: No. 206 Squadron collection*)

6 A U-boat under attack by Liberator 'K/206' (EV943) on 10 June 1944 after it had been damaged by four Mosquitos (*Photo: IWM. CH568*)

exercises in the Indian Ocean in 1954, the squadron undertook a 30,000-mile tour in seven weeks that took it as far afield as New Zealand and Fiji. Duties in connection with nuclear bomb trials at Christmas Island in 1956-57 took No. 206's Shackletons to Australia and the Pacific.

Some Shackleton MR.2s supplemented the MR.1s for a short period in 1953-54 but it was the tricycle-undercarriage M.R.3 which replaced the first of the tailwheel breed early in 1958. The first arrived on 14 January 1958; and, by the end of May, the old Mark 1s had been phased-out. In 1960, the squadron flew to Argentina and in the following year to South Africa. In July 1965, No. 206 moved to Kinloss.

Attempts by Indonesia to disrupt the new state of Malaysia included the sending of small raiding forces to the mainland and Borneo. For the first three months of 1966, the squadron deployed to Singapore to fly patrols over the Malacca Straits and off Borneo. Such worldwide operations extended the Shackletons to their limit and to provide extra power they were fitted with two Armstrong-Siddeley Viper jet engines built into the inboard engine nacelles.

No. 206 was presented with its squadron standard by HRH Princess Margaret on 28 July 1966. Detachments to the Middle and Far East continued and in February 1967, two of the squadron's Shackletons flew round the world. But, after 18 years, the end was in sight for No. 206's Shackletons. After conversion training at No. 236 OCU at St Mawgan, the squadron took over a complement of Hawker Siddeley Nimrod MR.1s, an advanced maritime reconnaissance aircraft developed from the de Havilland Comet pioneer jet airliner. With its sister squadrons, Nos. 120 and 201, it forms part of the maritime reconnaissance wing at Kinloss guarding the northern flank of NATO.

SQUADRON BASES

Ste Marie Cappel, France	1 April 1918
Boisdinghem, France	11 April 1918
Alquines, France	15 April 1918
Boisdinghem, France	29 May 1918
Alquines, France	5 June 1918
Ste Marie Cappel, France	5 October 1918
Linselles, France	24 October 1918
Nivelles, Belgium	26 November 1918
Bickendorf, Germany	20 December 1918
Maubeuge, France	27 May 1919
Alexandria, Egypt	19 June 1919
Heliopolis, Egypt	24 June 1919
Helwan, Egypt	27 June 1919 to 1 February 1920
Manston, Kent	15 June 1936
Bircham Newton, Norfolk	30 July 1936
St Eval, Cornwall	30 May 1941 (completed 9 July 1941)
Aldergrove, Co Antrim, Ulster	12 August 1941
Benbecula, Outer Hebrides	30 June 1942
Lagens, Azores	8 October 1943 (Ground echelon) 18 October 1943 (Aircraft)
Davidstow Moor, Cornwall	31 March 1944
St Eval, Cornwall	12 April 1944
Leuchars, Fifeshire	11 July 1944
Oakington, Cambridgeshire	1 August 1945 to 25 April 1946
Lyneham, Wiltshire	17 November 1947 to 20 February 1950
St Eval, Cornwall	27 September 1952
St Mawgan, Cornwall	14 January 1958
Kinloss, Morayshire	7 July 1965

Operational detachments:

Carew Chariton, Pembrokeshire	16 September 1939 to 10 October 1939
Hooton Park, Cheshire	10 October 1939 to 30 November 1939
Aldergrove, Co Antrim, Ulster	4 March 1941 to 18 April 1941
St Eval, Cornwall	6 May 1941 to 30 May 1941 5 December 1941 to 9 December 1941
Chivenor, Devonshire	11 February 1941 to 24 December 1941
Tain, Ross and Cromarty	25 May 1944 to 5 June 1944 30 July 1944 to 14 September 1944
Stornoway, Outer Hebrides	6 August 1944 to 23 August 1944

SQUADRON EQUIPMENT
Period of Use and Typical Serial and Code Letters

Aircraft	Period	Serial
Airco D.H.9	April 1918 to February 1920	D1689
Avro Anson I	June 1936 to June 1940	K6189 (VX-R)
Lockheed Hudson I	March 1940 to August 1942	N7318 (VX-W)
Lockheed Hudson II	April 1941 to August 1942	T9383 (VX-Q)
Lockheed Hudson III	April 1941 to August 1942	T9454 (VX-G)
Lockheed Hudson IV	April 1941 to August 1942	AE622 (VX-Y)
Lockheed Hudson V	October 1941 to August 1942	AM792 (VX-V)
Boeing Fortress II, IIA*	August 1942 to April 1944	FL460 (VX-V)
Consolidated Liberator VI	April 1944 to April 1945	BZ984 (PQ-S)
Consolidated Liberator VIII	March 1945 to April 1946	KK292 (PQ-P)
Avro York C.1	November 1947 to February 1950	MW303
Hawker Siddeley (Avro) Shackleton MR.1	September 1947 to May 1958	WG528 (B-U)
Hawker Siddeley (Avro) Shackleton MR.2	February 1953 to June 1954	WL742 (B-Z)
Hawker Siddeley (Avro) Shackleton MR.3	January 1958 to October 1970	WR980 (206-B)
Hawker Siddeley (D.H.) Nimrod MR.1	November 1970 to date	XV250

*Also a few Fortress Is employed for training purposes

COMMANDING OFFICERS

Sqn Cdr C. T. MacLaren	1 April 1918
Major G. R. M. 'Reid	24 June 1919
S/Ldr A. H. Love	15 June 1936
W/Cdr F. J. Vincent DFC	13 July 1936
W/Cdr H. O. Long DSO	22 October 1936
S/Ldr H. H. Martin	9 May 1938
S/Ldr N. H. d'Aeth	13 January 1939
W/Cdr J. Constable-Roberts	19 June 1940
W/Cdr C. D. Candy	20 February 1941
W/Cdr A. F. Hards	11 August 1941
W/Cdr H. D. Cooke	15 June 1942
W/Cdr J. R. S. Romanes DFC	29 June 1942
W/Cdr R. B. Thomson DSO	16 May 1943
W/Cdr A. de V. Leach DFC	23 March 1944
W/Cdr J. P. Selby	9 January 1945
W/Cdr T. W. T. McCombe OBE	24 July 1945
S/Ldr F. C. Blackmore	17 November 1947
S/Ldr J. C. Blair	15 March 1948
S/Ldr E. Moody	12 July 1948
S/Ldr E. A. Rockliffe	1 November 1949
S/Ldr J. D. Beresford	27 September 1952
S/Ldr E. K. Paine	8 December 1954
W/Cdr J. E. Preston	10 April 1956
W/Cdr R. T. Billett	14 April 1958
W/Cdr J. E. Bazalgette DFC	26 July 1960
W/Cdr D. R. Locke OBE	25 June 1962
W/Cdr H. R. Williams	27 October 1964
W/Cdr S. G. Nunn OBE DFC	15 August 1966
W/Cdr D. R. Dewar	1 October 1968
W/Cdr J. Wild	29 May 1970

Index

PERSONNEL

A

Abrams, W/Cdr. W. G. 43
Alcock, W/Cdr. J. M. 43
Annable, S/Ldr. K. 26
Armistead, W/Cdr. J. R. 57, 61, 63

B

Barber, S/Ldr. L. A. 26
Barker, Maj. W. G. 28
Barrett, W/Cdr. J. 38, 41, 43
Barrett, S/Ldr. K. J. 57
Bateson, W/Cdr. J.
Baveystock, Fl/Lt. 37
Bazalgette, W/Cdr. J. E. 84
Beaty, Fl/Lt. A. D. 80
Beauman, Fl/Lt. R. 17, 20
Bedford, S/Ldr. D. W. 43
Bentley, W/Cdr. G. W. 57
Beresford, S/Ldr. J. D. 84
Beringer, W/Cdr. W. 68
Berry, Sgt. 24
Berryman, P/O. 16
Betteridge, Cpl. G. 71
Biddell, Fl/Lt. 74
Billett, W/Cdr. R. T. 84
Blackmore, S/Ldr. F. C. 84
Blair, S/Ldr. J. C. 84
Blake, W/Cdr. E. A. 57
Booker, Maj. C. D. 28, 43
Boumphrey, S/Ldr. C. 45, 57
Bowry, S/Ldr. P. C. 26
Breadner, Maj. L. S.
Breakey, S/Ldr. J. D. 43
Brett, S/Ldr. D. E. 57

Britton, A.C.2. 73
Brown, W/Cdr. L. F. 57
Braithwaite, W/Cdr. F. J. St. G. 16, 20, 22, 26
Burn, Lt. 71
Burnett, W/Cdr. J. B. 43

C

Cahill, W/Cdr. C. H. 43
Camp, Sgt. 20, 21, 22
Campbell, F/O. K. 20
Campbell, S/Ldr. W. G. 14, 26
Candy, W/Cdr. C. D. 84
Carlisle, F/O. P. F. 81
Carr, S/Ldr. 74
Carrothers, Capt. 71
Case, W/Cdr. A. A. 57
Cecil Wright, S/Ldr. 33
Chesworth, W/Cdr. G. A. 43
Christian, Lt. L. A. 71
Clark, P/O. L. G. 77
Coates, W/Cdr. K. R. 43
Cohen, W/Cdr. L. 76
Constable-Roberts, W/Cdr. J. 84
Cooke, W/Cdr. H. D. 77, 84
Coote, W/Cdr. D. I. 68
Corballis, S/Ldr. E. R. L. 26
Coulson, W/Cdr. L. 57
Cowey, P/O. R. L. 77
Cox, S/Ldr. T. A. 57
Crosbie, W/Cdr. J. L. 43

D

d'Aeth, S/Ldr. N. H. 84
Dart, W/Cdr. A. D. 68

Davies, W/Cdr. A. C. 43
Davies, W/Cdr. E. S. C. 68
Davies, S/Ldr. L. W. 68
Davies, W/Cdr. W. A. L. 26
Davis, S/Ldr. M. J. 57
Deacon, S/Ldr. E. W. 57
Dewar, W/Cdr. D. R. 84
Dias, F/O. 73
Disney, W/Cdr. H. A. S. 43
Dobb, W/Cdr. B. E. 57
Donald, S/Ldr. D. G. 43
Dobell, F/O. 49
Duncombe, W/Cdr. J. J. 66
Dundas, Fl/Lt. A. D. S. 81
Dunn, S/Ldr. A. P. 26
Duxbury W/Cdr. J. B. 43

E

Eagleton, Fl/Lt. N. F. 53, 54
Ellison, S/Ldr. F. 57
England, S/Ldr. T. H. 26
Evens, S/Ldr. H. W. 57
Evison, W/Cdr. C. E. V. 68

F

Feather, S/Ldr. V. P. 68
Featherstone, P/O. 73, 74
Fenton, S/Ldr. 24
Finch, Fl/Lt. J. 54
Finlay, S/Ldr. D. O. 33
Fletcher, S/Ldr. L. A. W. 59, 68
Foster, Fl/Lt. K. H. 38
Frame, W/Cdr. A. 68
Francis, Fl/Lt. 20

85

G

Gayer, S/Ldr. W. A. 57
Gee, S/Ldr. R. 24, 26
Gerrard, S/Cdr. E. L. 44
Gibbs, Fl/Lt. P. 20
Gill, A.C.1. 73
Goble, A/Cdre. J. 59
Good, S/Ldr. A. G. A. 26
Goodwin, S/Ldr. E. S. 26
Gow, Maj. R. 57
Grannum, F/O. C. W. 48
Gray, P/O. 74
Greenhill, P/O. 73

H

Haggas, Fl/Lt. G. R. 81
Hall, Fl/Lt. 37
Hards, W/Cdr. A. F. 84
Harger, W/Cdr. G. P. 57
Harper, P/O. 73
Hartley, S/Ldr. H. S. 68
Hatfield, W/Cdr. P. R. 68
Hawkins, W/Cdr. H. J. L. 68
Hawroyd, F/Sgt. 23
Hearn-Phillips, Sgt. N. 20
Henderson, P/O. 73
Henry, S/Ldr. K. 57
Hillman, Sgt. 21
Horner, S/Ldr. D. H. F. 43
Horner, W/Cdr. T. Q. 53, 57
Howard, Capt. G. R. 11
Hughes, S/Ldr. 73
Huskisson, Sqn/Cdr. B. L. 68
Huskisson, Maj. P. 68
Hutton, Fl/Lt. D. R. 38
Hyde, Fl/Lt. E. L. 60, 61, 63

J

Jenkins, S/Ldr. H. H. 68
Jones, W/Cdr. J. H. O. 57, 43
Jones, W/Cdr. N. 43

K

Kane, Fl/Sgt. 24
Kean, P/O. 74
Kennan, P/O. 75
Kent, W/Cdr. P. 68
Kershaw, S/Ldr. R. H. 57
Kiste, W/Cdr. R. E. G. van der 43

L

Lander, W/Cdr. J. M. 24, 26
Laurence, S/Ldr. F. H. 59, 68
Leach, W/Cdr. A. de V. 81, 84
Learmount, Maj. L. W. 26

Leger, S/Ldr. R. J. M. St. 26
Leman, Maj. C. M. 28, 43
Leppard, S/Ldr. D. E. 68
Levien, Fl/Lt. 53
Lillie, Cpl. 61
Lindsay, W/Cdr. D. S. 54, 57
Lloyd, W/Cdr. C. K. N. 68
Lloyd, W/Cdr. K. B. 59, 68
Locke, W/Cdr. D. R. 84
Long, W/Cdr. H. O. 84
Longmore, Air Chief Marshal, Sir Arthur, 30, 42
Louw, W/Cdr. J. W. 43
Love, S/Ldr. A. H. 84
Lucas, Capt. the Hon. Lord 26
Lucy, Maj. R. S. 68

M

MacCallum, Fl/Lt. 52
McCombe, W/Cdr. T. W. T. 84
McCready, S/Ldr. R. A. N. 43
McKelvie, Maj. J. A. 26
McLachlan, F/O. 24
MacLaren, Sqn/Cdr. 70, 84
Mack, S/Ldr. R. E. X. 26
Maitland, S/Ldr. C. E. 26
Martin, S/Ldr. H. H. 84
Martin, Fl/Lt. L. G. 59
Martyn, Maj. R. B. 26
Mayhew, W/Cdr. J. C. 21, 26
Mellor, W/Cdr. H. V. 16, 26
Michell, W/Cdr. D. 68
Middleton, F/O. 33
Moody, S/Ldr. E.
Mullins, Sgt. 21
Mulock, Fl/Sub-Lt. R. H. 44

N

Nanson, Gp/Capt. E. R. C. 30
Nicholas, S/Ldr. C. H. 26
Nicholls, Fl/Lt. 53
Noakes, S/Ldr. J. 26
Northcott, W/Cdr. W. S.
Norton, Maj. E. W. 68
Nunn, W/Cdr. S. G. 84

P

Paget, Lt. 71
Paine, S/Ldr. E. K. 84
Parkhouse, S/Ldr. R. C. L. 43
Pegg, F/O. A. J. 13
Pegler, S/Ldr. R. A. 68
Percival, Lt. 71
Petheram, S/Ldr. C. J. 57
Phillips, Fl/Lt. 60, 61, 63
Pickles, W/Cdr. K. F. T. 68
Pitt, Lt. G. A. 71
Powell, Fl/Lt. 54
Preston, W/Cdr. J. E. 84

Q

Quinnell, Maj. J. C. 26

R

Reekie, S/Ldr. R. G. 57
Reeson, Sgt. J. 56
Reid, Maj. G. R. M. 84
Riccard, W/Cdr. C. S. 43
Ridgeway, S/Ldr. M. V. 26
Ritchie, S/Ldr. J. R. 26
Roache, W/Cdr. R. B. 43
Roberts, W/Cdr. J. G. 43
Robson, Maj. V. A. H. 26
Robinson, Capt. J. 57
Rockliffe, S/Ldr. E. A. 84
Roe, W/Cdr. R. D. 68
Rogers, W/Cdr. A. D. 57
Romanes, W/Cdr. J. R. S. 84
Roxburgh, Fl/Lt. W. 77
Rumbold, S/Ldr. P. A. S. 43
Ruth, S/Ldr. W. D. B. 37, 38
Rutherford, F/O. J. A. 48

S

Salter, S/Ldr. A. 26
Samuel, F/O. A. C. I. 77
Scott, Sgt. 21
Selby, W/Cdr. J. P. 84
Sheardown, Fl/Lt. H. R. 54
Shepherd, S/Ldr. P. A. 57
Sitwell, S/Ldr. W. G. 57
Smyth-Piggott, Fl/Cdr. J. R. 44
South, W/Cdr. P. G. 43
Spry, S/Ldr. B. A. 57
Stack, Fl/Lt. 37
Stafford, S/Ldr. G. 57

T

Taylor, S/Ldr. P. T. 57
Thain, S/Ldr. G. T. 57
Thomas, S/Ldr. 63
Thompson, Fl/Lt. G. B. H. 81
Thomson, W/Cdr. R. B. 77, 84
Thynne, F/O. B. W. 81
Todman, S/Ldr. D. A. W. 26
Townsend, L.A.C. 74
Tremear, W/Cdr. W. H. 43
Turner, S/Ldr. E. F. 43, 57

V

Vaughan, Fl/Lt. 33
Verran, S/Ldr. G. L. 26
Vincent, W/Cdr. F. J. 71, 84

W

Walshe, F/O. 54
Walters, Fl/Lt. I. F. B. 38
Wann, S/Ldr. A. H. 57
Ward, P/O. 75
Warne-Brown, S/Ldr. T. A. 14, 26
Warneford Fl/Sub-Lt. R. A. J. 28
Warren, Capt. H. 71
Weaver, S/Ldr. J. 26
Weller, W/Cdr. J. C. W. 68
Wemp, Sqn/Cdr. B. S. 57
Whitfield, Sgt. 77
White, P/O. 22
Wigglesworth, S/Ldr. C. G. 43
Wild, W/Cdr. J. 84
Wildy, W/Cdr. E. P. 68
Williams, W/Cdr. H. R. 84
Williams, W/Cdr. O. G. 68
Wood, S/Ldr. R. 57
Woollatt, F/O. 16

Y

Young, S/Ldr. G. 68

AIRCRAFT

BRITISH
P = indicates a photograph
CP = indicates a colour plate

Avro
500   44
503   44
504A   11, 26
Anson   15, 71, 72, 73, 73P, 74, 78CP, 84
Shackleton   35CP, 41, 41P, 42, 42P, 43, 43P, 65, 67CP, 68, 68P, 79CP, 82, 82P, 83P, 84

York   84

Blackburn
Dart   14
Ripon   48
Skua   52, 74
Turcock   13

Bristol
Beaufighter   19CP, 23P, 24, 26
Beaufort   14, 15, 16, 16P, 17, 17P, 19CP, 20, 21, 21P, 22, 23, 23P, 26, 63
Blenheim   15, 20, 22, 23, 63
F.2A; F.2B   12, 12P, 13, 13P, 18CP, 26
Scout   11, 26
Sycamore   25
T.B.8   44

de Havilland
D.H.4 (Airco)   44, 44P, 45P, 46CP, 57, 70
D.H.6 (Airco)   72
D.H.9 (Airco)   37, 70, 70P, 71, 72, 84

Dragon Rapide   72
Mosquito   24, 26, 56
Nimrod   See under 'Hawker Siddeley'
Tiger Moth   73

Fairey
IIID   45, 45P, 46CP, 48P, 57
IIIF   45, 46CP, 48, 48P, 49, 49P, 50P, 57
Albacore   22
Fulmar   54
Swordfish   17, 22, 52, 53, 57, 63

Gloster
Gamecock   13

Handley Page
Halifax   56, 57
Hampden   74, 80
Hastings   47CP, 56, 56P, 57, 82

Hawker
Horsley   14
Hunter   25

Hawker Siddeley
Nimrod   35CP, 42, 42P, 84
Shackleton   See under 'Avro'

Royal Aircraft Factory
F.E. 2a, F.E. 2b, F.E. 2c   11, 11P, 12, 18CP, 26

Saunders-Roe (Saro)
London   30, 31P, 32P, 33, 34CP, 43, 49, 50, 51, 51P, 52, 52P, 53, 53P, 57, 59, 62P, 68

Short
S.28   44
S.30   63
S.34   44
Calcutta   30
Shirl   14
Singapore   17, 50
Sunderland   33, 35CP, 36P, 37, 38, 39P, 40, 40P, 41, 43, 50, 51, 52, 54, 55, 56, 57, 60, 61, 62P, 63, 64, 64P, 65P, 66CP, 68, 68P, 75

Sopwith
Camel   27, 28, 34CP, 43, 45, 59, 68
Cuckoo   14, 15P
Pup   44, 59
Snipe   28, 28P, 43
1½ Strutter   12, 44, 59

Supermarine
Scapa   47CP, 49, 50, 50P, 51P, 57, 58P, 59, 60P, 61P, 68
Southampton   27P, 29P, 30, 31P, 34CP, 43, 48, 49, 58P, 59, 59P, 68
Walrus   24

Vickers
F.B.5   44
Valetta   65, 67CP, 68
Vildebeest   14, 25, 15P, 16, 16P, 18CP, 26

Vincent   14
Warwick   81
Wellesley   17
Wellington   22, 38, 64, 80

Westland
Whirlwind   19CP, 24P, 25, 26, 56, 57, 57P

AMERICAN

Boeing
Fortress   74, 76P, 77, 78CP, 80, 80P, 84
B-29 Superfortress   24

Consolidated
Catalina   24, 37, 40P, 47CP, 53, 53P, 54, 54P, 55, 55P, 56, 57, 57P, 64, 80, 80P, 81
Liberator   37, 56, 64, 77, 78CP, 80, 81, 82, 82P, 83P, 84

Curtiss
Hawk   53
J.N.3   11

Douglas
Dakota   40, 64, 67CP, 68, 82
Skymaster   40

Lockheed
Hudson   16, 22, 33, 54, 63, 73P, 74, 75, 75P, 76P, 77, 78CP, 84
Ventura   54

Martin
Maryland   17

North American
Harvard   24, 74
Mustang   24

GERMAN

Arado
Ar 196   63

Dornier
Do 18   73

Focke-Wulf
FW 200 *Condor*   30, 33, 37, 54

Heinkel
He 60   63
He 111   63
He 115   73, 74
He 177 *Greif*   37

Henschel
Hs 293   37

Junkers
Ju 88   23, 60, 61

Messerschmitt
Bf 109   22, 23, 63, 74
Bf 110   20, 63, 75, 81

87

UNITS

Squadron

1	RNAS 27
1	RFC 27
2	RNAS 44
4	RNAS 59
6	RNAS 69
6	RFC 70
10	RAAF 38, 75
11	RNAS 69
13	RAF 11
15	13
19	72
22	11 to 26 incl.
33	11
36	14
39	23
42	14, 15, 16
44	74
48	71
70	65
78	65
84	24, 65
86	22
89	24
95	64
98	45
100	14
114	65
172	38
177	24
190	80
201	27 to 43 incl.
202	44 to 57 incl.
203	29
204	58 to 68 incl.
205	65
206	69 to 84 incl.
210	45
211	24
216	65
217	24, 73
220	42, 56, 77, 80
228	51, 52, 56, 60
230	38, 40, 41
233	16, 45
240	33, 60
254	23
270	64
275	25
502	73
518	56
610	73

Operational Conversion Units

229	25
235	40
236	42, 84

Heavy Conversion Unit (HCU)

1674	80

Wing

100	33
295	64
346	24
901	24
907	24

SHIPS

British

Acasta *destroyer* 63
Brilliant *destroyer* 63
Cameronian *transport* 14
Centurion *target ship* 72
Curacao *light cruiser* 14
Courageous *aircraft carrier* 14
Dumana *base ship* 51, 52
Forester *destroyer* 52
Furious *aircraft carrier* 14, 15
Glorious *aircraft carrier* 48
Havelock *destroyer* 38
Hesperus *destroyer* 38
Hood *battlecruiser* 14, 20, 33
Icarus *destroyer* 60
Imperial *destroyer* 33
Isis *destroyer* 53
Kensington Court *merchant ship* 60
King George V *battleship* 20
Leander *light cruiser* 14
Manela *base ship* 59, 63
Nelson *battleship* 20
Ormonde *transport* 22
Ramillies *battleship* 14
Renown *battlecruiser* 14, 20
Repulse *battlecruiser* 20
Revenge *battleship* 14
Rodney *battleship* 20
St. Day *tug* 52
Snapper *submarine* 73
Velox *destroyer* 63
Wishart *destroyer* 52

German

Alsterufer *merchant ship* 37
Admiral Scheer *armoured ship* 33
Bismarck *battleship* 22, 33, 37, 75
Bremen *liner* 20
Deutschland *armoured ship* 60
Europa *liner* 20
Glucksberg *merchant ship* 52
Gneisenau *battlecruiser* 20, 21, 22
Prinz Eugen *heavy cruiser* 22, 37, 75
Scharnhorst *battlecruiser* 16, 20, 22, 63
Tirpitz *battleship* 22

Others

Alabastro *Italian submarine* 54
Australia *Australian heavy cruiser* 63
Durbo *Italian submarine* 53
Bilderdyke *Netherlands merchant ship* 60

MISCELLANEOUS CODE & OTHER NAMES

ASV radar—Air-to-Surface-Vessel radar 74
'Bircham Barrels' 77
Doraplane and Toraplane 17
Rover 22
Rhubarb 24
'Gardening' 16
'Vegetables' 16
Leigh Light 57
Lindholme gear 81
Sonobuoys 81

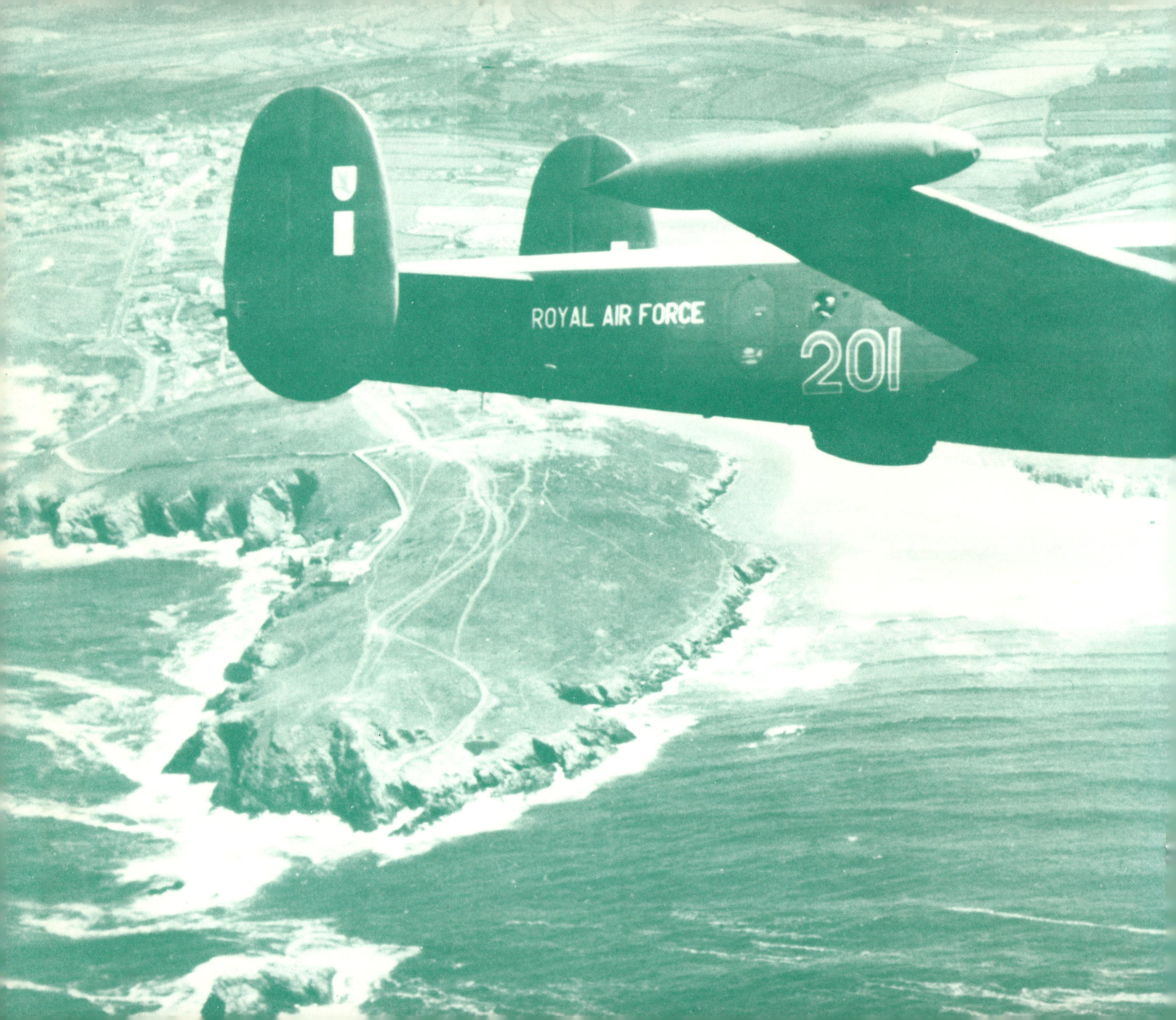
ROYAL AIR FORCE
201